STATISTICAL LITERACY

STATISTICAL LITERACY
A BEGINNER'S GUIDE

RHYS C. JONES

1 Oliver's Yard
55 City Road
London EC1Y 1SP

2455 Teller Road
Thousand Oaks
California 91320

Unit No 323-333, Third Floor, F-Block
International Trade Tower
Nehru Place, New Delhi – 110 019

8 Marina View Suite 43-053
Asia Square Tower 1
Singapore 018960

Editor: Charlotte Bush
Editorial assistant: Rhiannon Holt
Production editor: Martin Fox
Marketing manager: Ben Sherwood-Griffin
Cover design: Sheila Tong
Typeset by: C&M Digitals (P) Ltd, Chennai, India

Library of Congress Control Number: 2023942474

British Library Cataloguing in Publication data

A catalogue record for this book is available from the British Library

ISBN 978-1-5297-5480-3
ISBN 978-1-5297-5479-7 (pbk)

CONTENTS

ABOUT THE AUTHOR

Figure 0.1
Professor Rhys C. Jones

Rhys is Professor of Statistical Literacy, and an internationally recognised educational leader with extensive experience in curriculum development and curriculum theory, statistics education, and engaging students in small and large classroom settings (offline and online). He also has extensive experience of using digital literacy skills and learning analytics to enhance the student experience, using digital platforms to assess student engagement and interaction. Rhys was previously a Professor at the University of Surrey, the National Representative for the New Zealand Association for Gifted Children from 2020–2022, and he was also the Director of the Science Scholars at the University of Auckland from 2018–2021. Rhys also is a member of the Royal Statistical Societies teaching section as well as the Education Policy Advisory Group, and he is also a trustee of the Teaching Statistics Trust. These committees work at a national and international level to improve the teaching of statistics and data science in schools, universities, and relevant industry sectors.

Rhys was previously based in the Department of Statistics at the University of Auckland, which is a world-leading department in the areas of statistics and statistics education. He has a broad academic background in the areas of biology, chemistry, statistics, and education, and has held lecturing positions at the University of Surrey, University of Auckland, Cardiff University, London Southeast College, and Birmingham City University. Over his career he has taught a variety of subjects, at both undergraduate and postgraduate level, which include statistics, quantitative methods, mathematics for science, teacher training, research methods, biomedical science, nutrition and organic chemistry, health and well-being, and clinical anatomy and physiology. He has also taught a variety of science-based subjects in further education colleges, which include GCSE, BTEC, Access, and A levels. His primary research contributions are in the areas of curriculum development, randomness misconceptions, and the role of context.

Rhys has developed extensive experience in using active learning approaches in the teaching of statistics, to thousands of students over his career. In using these approaches, he has thoroughly enjoyed the challenge of bringing to life the teaching of statistics, across a range of disciplines and education levels (in schools and at university) across the planet. These teaching and learning activities have been incorporated throughout this book, to help engage and excite people in the art of statistical literacy, and the creation of compelling data stories.

ONLINE RESOURCES

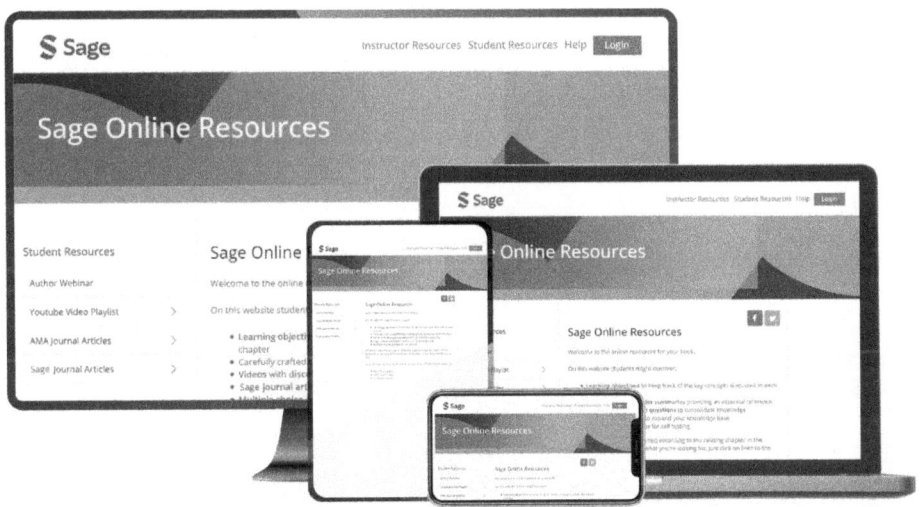

Statistical Literacy is accompanied by online resources for students. Find them at: **https://study.sagepub.com/jonesstatsliteracy**.

For Students

- Test yourself with **case studies and exercises** that help you put your learning into practice by analysing and evaluating examples and answering reflective questions. Each case study comes with answers and guidance so you can check you're on the right track.
- Practice your new data skills on the **datasets** discussed in the book. These are particularly relevant for Chapters 4 and 5.
- **Weblinks** to the real-world examples of data use (and misuse!) covered in the book mean you can easily follow along with the author's discussion.
- Use **multiple-choice questions** to check your understanding and get a sense of where you're confident and where you might need to go over a topic again. Answers to all the questions are provided for you.

1

INTRODUCTION

We are living in an increasingly data-centric world, one in which it is essential for everyone to have statistical awareness and develop appropriate statistical skills, which can include being able to communicate results from a statistical analysis, or present graphs in a report that helps to tell a convincing story or put across a point of view. The need for statistical expertise has never been greater. In rapidly changing times, and a world saturated with interesting data, we need new ways and new ideas to help make statistics more accessible for everyone. Statistical education has experienced seismic shifts in recent years, with respect to content and teaching methods. However, there is a lot further to go, particularly with the rise of data science which can, and should, transform the teaching, learning and practice of statistics.

Nothing illustrates the need for statistical and data science literacy more than the Covid-19 pandemic. Data were presented daily with respect to the number of cases, the number of hospitalisations, the number of deaths, and the number of vaccinations for different countries, regions and cities. Naturally, people want to see how their region compares with others or use it to make decisions about, for example, whether to get vaccinated, when to wear a mask, or where to travel. To make decisions like this, people need to be able to evaluate and interpret data in a critical but well-informed manner.

Statistical literacy includes the ability to understand and reason with data, skills that are increasingly important in making informed decisions related to both the world of work and personal decisions about health and family finances, but also societal issues such as climate change, inflation, migration, and the impacts of Covid-19. For example, those who refuse to get vaccinated against Covid-19, because they believe it is dangerous due to possible side-effects, often fail to consider that not getting the vaccination is not free of risk either. The more thoughtful response would be for the person to compare the probability and associated worst-case scenario of each. It is this type of statistical reasoning, a key component of statistical and data science literacy, that needs to be nurtured and developed in everyone.

An area that also needs consideration is the ongoing impacts of artificial intelligence (AI) and AI-assisted software on education, and statistics education. AI has had a transformative impact on statistics education, revolutionising the way we learn and apply statistical concepts. With the advent of AI, we now have access to powerful computational tools that can quickly analyse massive amounts of data, enabling the exploration of complex statistical models and the ability to draw meaningful insights. AI-powered platforms and software have also made statistics more interactive and engaging, offering simulations and visualisations that enhance understanding of abstract concepts. Additionally, AI has automated tedious tasks such as data cleaning and preprocessing, freeing up time to focus on higher-level thinking and problem-solving. However, as AI continues to evolve, it is crucial to emphasise the ethical considerations surrounding data privacy, bias, and the responsible use of AI in statistics, ensuring the development of a well-rounded understanding of both the benefits and limitations of these technologies. Many of these important topics will be touched upon throughout this book.

This book introduces many of the skills you will need to become statistically literate. As your confidence begins to build, in being able to produce engaging and clear data stories, you should also start to notice how much these skills support you every day in your own life. This could be in your own studies at school, college, or university, or perhaps in your place of work and associated activities.

1.1 Sensitive data

The world is full of data, which can often include controversial or sensitive topics. This section gives an overview of how this book will deal with data on topics such as gender diversity, sexuality, religion, ethnicity, race, politics, philosophical beliefs, and other topics related to the social sciences.

Figure 1.1 Sensitive data
Source: Photo by Malu Laker on Unsplash

1.1.1 What are sensitive data?

We live in a world full of exciting and engaging subjects that are all underpinned by data. Data can come in many different forms and can be used in a variety of contexts.

The same data can also be used to explore many different contexts. Context and subject-matter can give data a new lease of life, used to convey a message or set of ideas. Different contexts and subjects can mean different things to different people, both on an individual level and to different cross-sections of society. Naturally, the use of sensitive data sets can be linked to controversial or sensitive topics, which certain groups of people may want to avoid. We therefore need a definition of what sensitive data sets are, to explain and highlight when and where they are going to be used in this book, to enable the reader to decide whether they would rather avoid such examples.

To define sensitive data is not easy, since such a definition can encompass so many kinds of data, across a

> **Sensitive data** is defined as any data set that is linked to a controversial or potentially sensitive topic, that can be emotionally triggering, upsetting or cause distress.

multitude of contexts and topics. It therefore feels appropriate to assign a broad definition to how this book will flag sensitive data.

1.1.2 Why deal with controversial or sensitive topics?

So many controversial or sensitive topics will often involve cross-sections of society, which can come from different cultures. They can describe or explain the way we live, how people identify themselves, and they can also touch on social norms and beliefs. They can define the way we live and interact with others. It is therefore *essential* that we collect and use data on these topics to better understand ourselves and society. To censor and exclude such topics distorts the truth and can give a false image of the world we live in.

As an educator, I believe that it is my responsibility to show my students the real world, warts and all. I feel I have a moral obligation to get students talking about sensitive topics, or issues that are perceived to be controversial. In doing so, I feel that I can better prepare my students for the world they live in, and for when they enter the world of work and beyond.

Learning that data are important to help us explain the world around us can support the development of a more rounded understanding of who we are, our own position in it, and can help us to challenge our own beliefs and ideologies. It can also foster a deeper understanding of other people's perspectives, make us more empathetic, and lead us to delve deeper into epistemological and ontological considerations. It is also a great way to develop one's own critical thinking skills.

Critical thinking is a vital skill that empowers individuals to analyse, evaluate and interpret information effectively. It goes beyond accepting ideas at face value and encourages a deeper understanding of complex issues. The importance of critical thinking lies in its ability to foster independent thought, logical reasoning and sound decision-making. By questioning assumptions, examining evidence and considering alternative perspectives, critical thinkers can arrive at well-informed conclusions and avoid falling prey to biases or misinformation. In today's information age, where we are constantly bombarded with vast amounts of data and opinions, the ability to think critically allows us to sift through the noise, distinguish fact from fiction, and make thoughtful choices. It equips us with the tools necessary to solve problems creatively, adapt to new challenges, and navigate an increasingly complex and interconnected world. Furthermore, critical thinking promotes intellectual curiosity, empathy, and open-mindedness, enabling constructive dialogue and fostering a more inclusive and democratic society.

I do recognise, however, that certain topics can be emotionally triggering, upsetting and too challenging for some. Which takes us back to moral obligations and duty of care. As an educator I also have a responsibility to ensure my learners feel safe and comfortable with the content they are learning. Getting the balance right is not always easy, but this section is an attempt to help you make your own decisions with what you feel comfortable learning (and, for instructors, what you feel comfortable with teaching, and what you would like your students to learn, as well as topics you would rather they avoided).

1.1.3 How this book deals with sensitive data

Throughout this book, you will come across a multitude of data sets, reports and data displays. When topics or questions/tasks include sensitive data, or topics that are likely to be controversial or potentially sensitive to the reader, they will be clearly highlighted as 'SENSITIVE DATA'. There will also be a short note of the title of the topic, which could include sexuality, gender diversity or religion. This will make it easier for you to decide whether you would like to avoid the section identified as containing sensitive data.

1.1.4 Framework for dealing with sensitive data

If you do decide to engage with sections of the book identified as 'SENSITIVE DATA', this section will help you to navigate those sections safely, ensuring you can maximise your learning opportunities.

Use the following prompts to help you navigate 'SENSITIVE DATA' sections:

1　　What is the general heading assigned to the 'SENSITIVE DATA'? For example, is it to do with gender diversity? Or religion? Ask yourself how this topic makes you feel. Does it mean something personal to you? At this point, if you feel it is emotionally triggering or too sensitive a topic for you to manage, it might be better for you to skip it.

2　　If you feel comfortable with the topic, next ask yourself why you think it has been chosen. Try to think about what it can tell us about the world we live in. Is it interesting? What issue is it tackling or getting you to think about? Your tutors and lecture might already show you bad examples of a report or piece of work relevant to what you are learning. And there is no shortage of these in the world!

Engaging with exercises like this helps you to develop your critical thinking skills and can also nurture your abilities to distinguish between what is deemed to be good, and what is not so good.

3   When looking at the data, think about why they were collected, and how they are being used in the report or story. What is the agenda of the author of the report or story you are looking at, and can this be used to explain their perspectives or views being portrayed?

4   Finally, think about the intended audience of the report or story. This can really help you to understand the language a piece of work uses to deliver its key messages. It can also help you to unearth the reasons why the authors have taken a certain angle, and whether you think they have a case, that is, how persuasive is the piece of work being presented.

5   Critiquing other people's work is often a lot easier than creating your own! Examples chosen in the 'SENSITIVE DATA' sections will often ask you to create your own version of the story or report being displayed. These can be from your own perspectives, or you might be asked to write a report, using data, from the perspective of somebody else. Bear this is mind when reviewing 'SENSITIVE DATA' examples.

1.2 The structure of the book

The first half of the book (Chapters 2–6) introduces you to the essential skills needed to become statistically literate, with each of the corresponding chapters comprising of a series of 'Develop your skills' activities. Suggested answers are included at the back of the book and online.

Chapter 2 ('Building a Foundation') will introduce you to the importance of using good English and grammar when communicating information related to data. It will also serve as a refresher to what makes a good sentence, in terms of creating a clear and coherent narrative. This chapter will set the scene for the whole book.

Chapter 3 ('Polls and Surveys') will focus specifically on the ways in which polls and surveys generate and then communicate data. Examples of real polls and surveys will be presented to the reader (e.g., via platforms like SurveyMonkey and Qualtrics), outlining the diversity in these tools for collecting data. An exploration of the language used in them will be undertaken, along with a focus on the reasons behind their creation and uses. Sampling and non-sampling errors will also be reviewed briefly. You will be guided through a series of tasks to reinforce elements of what makes polls and surveys good and bad.

Chapter 4 ('Introduction to Categorising and Visualising Data') will dispel the myth that statistics is the same as mathematics. It will include a series of examples demonstrating how uncertain this discipline can be. Real-world data sets will be used, presenting guidance on how to display data, in multiple ways. This chapter will focus on the different types of variables that exist, along with the ways in which they can be displayed. Through guided tasks, you will recognise the many shades of grey that exist within statistics. This will then link well with the next chapter, which looks at some of the more specialised terms used in statistics.

Chapter 5 ('The Language of Statistics and Statistical Inference') will build on the previous chapter, with emphasis placed on key words that are used in presenting and communicating interesting patterns and trends in data, as well as the results from statistical analyses that may have been undertaken. You will be encouraged to see the distinction between using certain words in statistics, reinforcing the shades of grey analogy in Chapter 4. This chapter will help you to develop your statistical literacy skills, using a range of real-world examples. Examples will be drawn from scenarios that involve the interpretation of confidence intervals, an introduction to statistical inference, and an overview of the difference between practical and statistical significance.

Chapter 6 ('An Introduction to Data Stories') will set the scene for why we need data, whether it is to answer a research question, create new knowledge, or to create an engaging and convincing story. This chapter will briefly touch on the ontology and epistemology of knowledge, the scientific method, and what makes good research questions. There will also activities to help you to explore the value of research and good storytelling.

In later chapters (Chapters 7–13), 'Develop Your Skills' activities are introduced within each chapter, to help you to apply skills developed in the preceding chapter of the book. These tasks often require you to think critically, and they will also help to develop your communication skills, as well as your statistical literacy confidence. Suggested answers are included at the back of the book and online.

Chapter 7 ('Media Reports and Social Media Platforms') will highlight that the use of data in the media and on social media platforms is ubiquitous. This chapter will draw on a range of examples, from the media and social media, emphasising the use of language to present a certain point of view. It will build on Chapters 3–6, reinforcing the essential statistical literacy skills and critical thinking skills covered in them. Tasks will be built around the examples presented, to enable the development of a greater depth of understanding as to how the media use language (and the potential biases they have) when communicating data.

Chapter 8 ('Experiments and Observational Methods in Research') will describe experiments and observational methods in more detail, discussing how different disciplines use them. It will draw on a range of examples, along with the language used to describe the many steps in them. Ethical considerations will also be explored, in relation to the methods selected in a study. Questions will be embedded throughout the chapter, drawing on well-known and interesting examples of experiments and observational studies, from a variety of disciplines (mostly psychology and social sciences).

Chapter 9 ('How to Read Journals with Quantitative Data') will begin to focus on data displays and information that you are likely encounter in your degree studies. Emphasis will be placed on the communication of data analysis and results in journal articles, including the language used in inferential statistics (introduced in Chapter 3). Articles based on experiments and observational studies will be included, with associated questions embedded around them, linking directly to the next chapter.

Chapter 10 ('Telling Stories with Descriptive Data, Tables, and Graphs') will provide a rationale as to why data stories are useful, as well as guidance for presenting data stories, including the importance of knowing the intended audience. The advantages of using descriptive data are also discussed, along with additional guidance about using graphs and tables in data stories.

Chapter 11 ('Common Misconceptions in Statistics and Statistical Literacy') will link directly to the previous chapters in the book (especially the previous two), presenting you with an overview of the common misconceptions in statistics. These misconceptions will be highlighted using a range of examples.

Chapter 12 ('Statistics Communicated Badly') will build on the previous three chapters, and Chapters 4 and 5, highlighting examples of statistical analysis and their communication done badly. This chapter will also give an overview of how to use graphs and tables to present data clearly, so as to create a coherent narrative (building on Chapter 3). You will be guided and encouraged to create your own short reports, based on data that you have collected.

Chapter 13 ('The Power of Statistical Literacy and Statistical Reasoning') will review the essential statistical literacy skills covered throughout this book, as well as introducing you to several new skills, that will help you with presenting and communicating statistics. This will include encouraging you to present your own critical arguments, and well thought-out data displays, to produce statistically sound reports.

Throughout the book, some of the 'Develop your skills' sections will include sensitive data examples. As discussed above, 'sensitive data' is defined as any data set that is linked to a controversial or potentially sensitive topic, that can be emotionally triggering, upsetting or cause distress. When these are included, it will be clearly signposted, so that you can decide whether you feel comfortable to engage with the relevant examples presented. There is also a framework, presented above, which gives you tips on how to deal with sensitive data, based on developing your critical thinking skills.

Throughout this book, an initial capital letter is used to denote a variable and also the levels in each variable. This has been done to enable you to spot variables and levels of measurement for those variables more quickly and easily. Don't worry if you don't know what a variable is yet – we will find out in Chapter 3. (For those who are curious: a variable is a characteristic that can be measured and have different values. Height, age, income, province or country of birth, grades obtained at school and type of housing are all examples of variables.)

There also resources to support each chapter, which include weblinks and links to other resources, to help support your learning within each chapter. For some chapters there are also additional references, for you to read further and expand your knowledge and confidence with the associated content. These links are also provided in the accompanying online resources, associated with this book.

I wish you well and hope that you enjoy the data journey you are about to embark on!

2
BUILDING A FOUNDATION

An overview of this chapter...

Given the title of this book, you might be expecting us to jump right into talking about numbers. But we're going to start with the 'literacy' part of statistical literacy. If you're wondering why – well, statistical literacy involves many of the raw ingredients that come with general literacy, which are fundamental in being able to clearly articulate the message you want to convey.

Literacy includes the ability to read and write. Writing a sentence can be achieved by many people and is something we learn and practise throughout primary and into secondary school. The ability to write a sentence and present it to others forms the basis of a large part of how we communicate with each other, whether it is done by writing a letter or via email or on other social media platforms such as Facebook or Instagram. The ability to communicate ideas and information in a sentence is an essential part of everyday life, for everyone! And with that comes the ability to read what has been written. However, what has been written, the way it has been formed, developed, and presented and its contents and clarity will all have a profound impact on the reader's ability to understand what they have read.

Statistical literacy involves many of these raw ingredients and is fundamental in being able to clearly articulate the message you want to convey. But the information being conveyed will have some sort of statistical or quantitative elements to it, perhaps involving numbers presented in some form. These skills are so important in being able to present and convince others of your arguments. This is why this chapter is one of the first in this book, representing a key golden thread that will weave into a glorious pattern as you progress through subsequent chapters.

Resources to support this chapter...

This chapter draws on a range of sources to help show you what good and bad sentences look like, using a range of real-world examples. You can build your skills by trying the suggested activities for constructing clear and coherent sentences to convey statistical or numerical information.

Resource	Date accessed	Location
iPad Air, *Apple*, 2021	23 July 2021	https://www.apple.com/uk/ipad-air/
Chelsea Clinton urges global sharing of COVID vaccine technology, *Nature*, 2021	23 July 2021	https://www.nature.com/articles/d41586-021-02164-8
Grammarly, 2021	23 July 2021	https://www.grammarly.com/

The websites referred to in all the activities are provided as part of this book's online resources. You can find them at **https://study.sagepub.com/jonesstatsliteracy**.

2.1 What makes a good sentence?

Being able to write well requires a lot of practice! And it's something you will be told repeatedly, especially by your tutors and lecturers. The more you practise, the better you'll become, and it's something you'll notice as well (especially when your grades get better).

Fortunately, you can practise constructing 'good' sentences. There are many ingredients that make a good sentence, and some are undoubtedly contested or a matter of preference. But using a list like the following – which is a guide and not comprehensive – can help you become more confident that you're writing in an informative, effective way.

The cardinal features of a good sentence (usually in academic writing styles, such as in an essay) often include the following:

1   *Limited content.* A sentence should include information that relates to one or two ideas. When you start to include lots of different pieces of information, the reader can become confused and lost.
2   *Achieving clarity.* Being able to write clearly does not mean you have to write an overly complex and long sentence. Using words that are clear and easily understandable will help you reach a wider audience.
3   *Being concise.* If your sentences are too long, the reader can get lost in them. Usually, one to three lines will be long enough for a sentence.
4   *Good grammar.* Using grammar well will ensure the verbs and nouns you use are in the right order, which can help to convey your message better.
5   *Good punctuation.* Being able to punctuate properly helps to add flesh to the bones of your sentence. It can help to add emotion and give your sentences different dimensions.
6   *Accuracy.* Ensure the information you include in your sentence is accurate, and preferably supported by others (i.e., reference other people's work in yours). This will be covered more in later chapters and will feature prominently throughout the rest of this book.

A useful way to check on the above is to read aloud what you have written. Ask yourself, does it make sense? Is your message being conveyed easily? Is it clear and concise?

Could it be improved? You could also give it to someone else to read and ask them what it means. See what they say and if it matches with what you intended for it to convey.

Here are two examples of well-written sentences, with an explanation outlining why they are good:

Example 2.1: Sample iPad Air advert

Powerful. Colourful. Wonderful.

iPad Air does more than a computer in simpler, more magical ways. And now with more features and capabilities, it is more versatile than ever.

Why these sentences are good:

- The opening three words are short, snappy and eye-catching.
- The content is simple, easy to read and engaging.
- Overall, the sentences are clear and concise.

Example 2.2: Nature news Q&A (9 August 2021)

(Weblink: https://www.nature.com/articles/d41586-021-02164-8)

Chelsea Clinton urges global sharing of COVID vaccine technology

The health-policy specialist who grew up in the White House is using her training and connections to convince world leaders to help make vaccines accessible to all nations.

Why these sentences are good:

- The initial headline is clear, concise and informative.
- The first sentence (starting with the words 'The health-policy') is informative and clear.
- There is good use of grammar and punctuation (good use of capital letters and full stops).

Try to avoid:

1   Starting your sentences with these words: for, nor, but, or, yet, so. Especially if you are writing formally.
2   Using complex words that you do not understand or cannot define without using a dictionary.
3   Making spelling mistakes. Spell-checkers have become very efficient; however, they are not foolproof. For example, they cannot always check the words in your

sentence are the ones you intended to use. For example, the word 'for' could have been accidently typed as 'far'. The spell-checker would not know that 'far' is not the word you wanted to use in the sentence you have written. Also check any specialised words you have used that are not in your spell-checker word bank. It may have changed one of your words into another one. A good way to help you check for this is to proofread your work and get others to double-check it for you.

······· Develop your skills! 2.1 ······································· ·····

Practise your skills with writing a 'good' sentence. Use the 6 points listed on p. 10 to help you with these tasks. See if you can change each of the following sentences to improve its clarity and conciseness:

1   There were lots and lots and lots of eggshells all over the very big floor in the kitchen by the toilet, across the way from the living room.
2   The most favourite sandwich in Burger King is a Whopper, then it's a chicken royale, then it's a cheeseburger.
3   Most people seem to very much like to at 9 a.m., run for, on average, about 34.5 minutes, then have breakfast after, in many suburbs in Auckland, which is in New Zealand, on the North Island.
4   Most people with a virus who contract a virus and get ill from it will recover, in many countries across the globe.
5   The results from an observational study proved that 54% of people lost so much of their concentration in the workplace, mostly males were affected, when they watched a lot of TV, which means around 54.32% of their spare time, which is about 5 hours of this.

2.2 Linking sentences and using paragraphs

Sentences are often grouped together into paragraphs. Each sentence should give you information about one idea. A group of sentences are then put together to form a paragraph, which usually includes information about one topic or subject. The sentences within a paragraph should link together and not jump around to other topics or ideas that are not connected. This enables the reader to follow the information being presented. Also think carefully about how you paragraphs link together, how they flow onto to each other and how they should be arranged into different sections or chapters.

This section will present several of the key ingredients needed to link sentences together. In the 'Develop your skills' exercise, you may have split several of the sentences up into smaller sentences, because there was too much information in them to begin with. This section will provide additional guidance on linking sentences together when you are writing sections or paragraphs.

Try to avoid:

1   Starting a paragraph with information that has not been introduced well or seems
     to appear from out of the sky!
2   Including too much information in one paragraph. As a rule of thumb, paragraphs
     should have between three and seven sentences in them.
3   Try to avoid writing about unconnected ideas in one paragraph. Ideally, they
     should contain information related to one topic or theme.

····· ·· Develop your skills! 2.2 ··· ···

Practise your skills with linking sentences together. You will need to rearrange them into an
order that logically flows and think about how the sentences link together. Use the points and
information above to help guide you in this task:

1   People from lower socioeconomic groups are more likely to lose their jobs and are more
     likely to use their savings to survive. During times of economic stress, people from different
     socioeconomic groups, in Western society, react differently. People in the Middle East will
     react differently. Whereas people from higher socioeconomic groups are more likely to invest
     their savings and top up their pensions. In conclusion, the rich get richer, whereas the poor
     get poorer.
2   Evidence-based approaches should be used instead. Learning involves a combination of
     cognitive abilities and life experiences. Many learning theories are contentious, and some
     are underpinned by weak empirical evidence. A more nuanced approach in using learning
     theories in relation to teaching practices, such as Bloom's taxonomy, should be adopted.
     Despite this, educators often take learning theories and use them as if they are facts,
     embedding them in the teaching methods and guidance developed.
3   Sadly, the field is also littered with unsubstantiated claims that have little or no evidence.
     Nutritional biochemistry can tell us useful things about the food and drink we consume. An
     interdisciplinary approach continues to aid and contribute to the important evidence base of this
     discipline. As a disciplinary field, the evidence used to support the knowledge base has been
     increasingly steadily over the last few decades.
4   Communicating verbally involves several key aspects that involve several human senses.
     For example, people's concentration span has decreased over the last decade, as well as
     their ability to verbally communicate with others. Social media increasingly influences the
     way many people live their lives. Many scientists claim that this is having a detrimental
     effect on human behaviour. Data show that people are spending a significant amount
     of time using social media platforms on their smart phones or other digital devices.
     Proponents of social media state that there are many benefits for people engaging with
     the multiple platforms available, including the creation of new and exciting forms of
     communication.
···

2.3 Using numerical data and numbers in a sentence

The final section of this chapter covers the key points involved in using numerical data, and numbers, in a sentence. When you have become confident with writing sentences – perhaps you are at that point now – the ability to insert numerical or statistical data is an important skill that you need to master. Research documents, government reports, news reports, even social media entries, often start with some sort of statistical or numerical data to help engage the reader. It helps to convey the magnitude or scale of what the writer wants you to become aware of. It can also be a way of setting the writer up to start talking about a solution or potential answer to a problem. This section will be built upon in several areas across the book, so think of this as a seed section, that will grow and flourish as you progress through other chapters of the book.

We have looked at how to write a 'good' sentence, and how to link them together to form a paragraph of connected ideas, covering one theme or topic. Now we will look at how numerical data can be used effectively in a sentence contained within a sentence or paragraph.

Statistical or numerical data can be used to convey the magnitude of a phenomenon or issue central to the information conveyed within a paragraph. They can also be used in standalone sentences, depending on the format of the information being presented (i.e., you might want to write an eye-catching heading in a journalistic piece of writing).

Percentages and proportions are essential concepts used in various fields, from mathematics and finance to statistics and everyday life. **Percentages** represent a proportion or fraction of a whole, expressed as a number out of 100. They allow us to compare and analyse data in a standardised way, making it easier to understand and interpret information. **Proportions**, on the other hand, refer to the relationship between different parts of a whole or the distribution of a quantity. They help us grasp the relative sizes and quantities of different components, enabling us to make informed decisions and draw meaningful conclusions. Whether we're calculating discounts, analysing statistical data, or evaluating the composition of a mixture, percentages and proportions play a fundamental role in quantifying and comprehending the world around us.

Percentages or proportions are often used to convey how far-reaching the issue or phenomenon you are writing about could be. They set the scene for the paragraph or piece of information you are writing about, and often act to engage the reader. These sentences need to contain accurate statistical information that links the numerical data with the subject-matter. This should be done clearly and not be presented in an ambiguous or confusing way.

Paragraphs often start with sentences like the ones describe above. However, they can also appear lower down in a paragraph, depending on the context being described.

Averages (usually the mean, which is calculated by adding together all values in a sample and dividing by the number of values in that sample) are also common pieces of statistical information that are communicated within sentences. They can appear in paragraphs, potentially to introduce the topic or theme being described. As with percentages and proportions, averages should be calculated or taken from other sources accurately, and they should link to the subject-matter well.

Here are some general rules when using numbers in a sentence (your university, or organisation, may also have similar guidelines to the ones presented here):

1   Spell out numbers that begin a sentence: 'Twenty-seven students received the Gold Axe Award.'
2   Spell out numbers used in a casual sense: 'I told you a hundred times to stop biting your nails.'
3   Use numerals for numbers 10 and greater and spell out numbers one to nine, with these exceptions:

    o   Addresses: 3 Knoles Dr.
    o   Ages, for people and objects: 2-year-old boy, 1-year-old book
    o   Dates: 8 January
    o   Dimensions: 5 feet high, 4-by-9 inches
    o   Highways or motorways: Route 5, M25
    o   Millions, billions: 6 million students
    o   Money: 5 pence, $7
    o   Percentages: 5%
    o   Temperatures: 9 degrees
    o   Times: 9 a.m.

Try to avoid:

1   Using percentages for small sample or population sizes (i.e., below 50). Use a proportion instead (i.e., 15 out of 20, which can be presented as 15/20). It is often useful to include the proportion or sample size in brackets, just after presenting the percentage in a sentence.
2   Presenting numbers to an inappropriate number of decimal places or **significant figures**. (As a reminder, decimal places are positions of the digits to the right of a decimal point. Rounding to, say three decimal places involves eliminating the digits after the third decimal place. To do this we look at the digit in the fourth decimal place. If that digit is 5 or higher we round the third decimal place up, that is, increase it by 1. If the fourth decimal place digit is 4 or lower we leave the third decimal place digit alone. Significant figures are the number of digits in a value, often a measurement, that contribute to the degree of accuracy of the value. We start counting significant figures at the first non-zero digit.) Too many decimal places or significant figures might make the reader have to work harder to read the numbers being presented and might confuse them. For most sentences, reporting a percentage to two decimal places should be sufficient.

Develop your skills! 2.3

Practise your skills with using numerical data in sentences. Correct the following sentences, which included a variety of numbers and data.

1   The temperature reached nine degrees by noon.
2   Our afternoon meal cost a total of fifty-five pounds and sixty-two pence.
3   We often liked to travel along the Mfour [roads in the UK are given a letter then number after, like the M8], which took us to Oxford.
4   There are approximately 23,000,000,000 results, when you search for the word 'number' in Google. The search takes nought point seven seconds.
5   It started snowing on February the ninth and lasted thirty-five days.

Develop your skills! 2.4

Practise your skills with using numerical data in sentences. Try to tidy up the following sentences to make them clearer and easier to read. Use the points and information above (in Section 2.3) to help guide you in this task:

1   87% of British Facebook users have admitted to looking at their ex-partner's profile, within two weeks of the break-up. 69% of these have also stated that they were the ones who were dumped by their ex-partner.
2   The most frequent day for people in the UK to order a takeaway is Friday, a recent poll by Readers Eating suggests. More than 10,000 people responded to the poll, mostly situated in the capital, London. Around 67% said Friday was their most popular day to order takeout, with Monday being the least likely day.
3   More than 52% of British adults have stated that they are worried about ageing. Of these, 62% were female, 22% were male and the rest were non-binary.
4   Approximately 7 out of 10 people in Grimsby use music platforms to stream or download music, a new study has found. The most popular platforms used were Spotify and Apple Music.

Develop your skills! 2.5

This exercise will help you to develop your critical thinking skills. There are many applications available online that can assist people with their writing skills, such as ChatGPT (https://openai.com/blog/chatgpt) and Grammarly (https://www.grammarly.com/). Answer the following questions related to Grammarly:

1   Outline the claims that the Grammarly application makes on assisting people with their writing.
2   What assumptions do you think the application producers at Grammarly make about the writing skills of people who opt to use it?

3   Do you think there are any limitations in using applications like Grammarly?
4   What are the long-term dangers of people using applications like Grammarly? What impact will
      it have on their writing skills?

. . . .

Key points to remember

The ability to write a good sentence is a fundamental skill needed to communicate and convey
information. Improving your skills in this area will take time but can be achieved with practice
and following the guidance and quizzes provided in this chapter. Below are a series of pointers to make sure you avoid common errors related to this topic:

1   Avoid writing long sentences (more than about three lines).
2   Do not use words that you cannot define or are unsure about. Avoid using overly complex
      language. Of course, you could aim to expand your vocabulary and learn new words. This is
      a good strategy to adopt, whereby you should practise using them in sentences and ensure
      you use them in the right context. But remember that clarity is better than coming across as too
      sophisticated or overly complex in your writing.
3   Try not to throw random sentences together and hope they form a coherent narrative. Think
      about the path you are taking your readers on and try to assess whether it has a logical flow.
4   Think carefully about the key messages you want convey in your paragraphs and sentences.
      Each word you choose to sit within your sentences should be thought out.
5   Read through what you have written and be as critical as possible. Ask yourself if it can be
      improved, think about whether the content makes sense, and try to assess if there are any
      ambiguities present.
6   Try giving your written work to someone else to read through, like a study buddy. They can be
      familiar with the subject-matter, or unfamiliar. Ask them to tell you what they think the main
      points are.

References to support this chapter

Apple (2021) iPad air. https://www.apple.com/uk/ipad-air/ (accessed 23 July 2021).
Maxmen, A. (2021) Chelsea Clinton Urges Global Sharing of COVID Vaccine Technology. *Nature*, 596
      (7872), 331–331. https://doi.org/10.1038/d41586-021-02164-8.

3

POLLS AND SURVEYS

An overview of this chapter

This chapter will delve into the exciting and increasingly expansive world of polls and surveys, which are created for many different reasons and come in many forms. Concepts covering sampling and biases with the data collected from polls and surveys will also be explored in this chapter, which will help you get to grips with subsequent chapters in the book (Chapters 4 and 5).

We are increasingly exposed to different ways of data being extracted from us, especially in digital form. Data collected from responses to a survey can be used to tailor emails that present you with specific offers from a retail store. Polls can be used to help predict the outcome of upcoming events, such as political elections. Polls tend to be based on one specific question, whereas surveys are larger, often including multiple questions and question types. In both forms of data collection, the results are often presented as percentages, proportions, and mean values. There are also other considerations to reflect upon when evaluating the usefulness of polls and surveys, for example their validity and reliability – can the results be generalised to larger populations? This chapter will address these areas, as well as looking at the different formats available and the language used in them.

Resources to support this chapter

This chapter draws on a range of sources to help show you get a better understanding of what polls and surveys are, and their value. In this chapter, we use a range of real-world examples to show you what good and bad polls and surveys look like, as well as delving deeper into concepts covering sampling and potential biases with associated data. You can build your skills by trying the suggested activities for looking into what makes a good poll or survey and investigating the language used by them.

Resource	Date accessed	Location
How to create a poll in 3 simple steps, *SurveyMonkey*, 2021	11 March 2021	https://www.surveymonkey.co.uk/mp/online-polls/
How to increase survey response rates, *Qualtrics* 2021	11 March 2021	https://www.qualtrics.com/uk/experience-management/research/improve-survey-response

(Continued)

(Continued)

Resource	Date accessed	Location
Skills assessment, *National Careers Service*, 2022	12 August 2022	https://nationalcareers.service.gov.uk/skills-assessment
Fragrance profiling, *Penhaligon's*, 2022	12 August 2022	https://www.penhaligons.com/uk/en/fragrance-profiling

The websites referred to in all the activities are provided as part of this book's online resources. You can find them at **https://study.sagepub.com/jonesstatsliteracy**.

....

3.1 A world full of polls and surveys!

Polls and surveys come in many different formats and styles. A poll is often much shorter than a survey, and usually includes only one question or statement that requires a response. Surveys tend to be much longer, and include a series of questions and statements that require multiple responses from the end user. Both methods ask questions to help ascertain the respondent's views on the topic of interest. The way they are constructed and subsequently delivered can have big implications on the response rate (i.e., how many people respond to the poll or survey) and the completion rate (i.e., how many people finish the poll or survey). While online formats are becoming increasingly popular (particularly in the aftermath of the Covid world we are currently living in), in-person and postal mail forms still give a sufficiently high of response rate. Having a good response rate (which most agree is usually above 50%) is important, since it can affect the validity of any conclusions or generalisations made about the data. For example, if there was a low response rate, this would make it difficult to extend findings to the wider **population** the sample was taken from.

Polls and surveys are often conducted to tell us something about one or more **variables** of interest. To better help you understand what a variable is, it can be defined as a characteristic that can be measured and that can have different values. Height, age, income, province or country of birth, grades obtained at school and type of housing are all examples of variables. If we were looking at a variable like grades obtained, A and B grades would be examples of data for that variable. Variables may be classified into two main categories: **categorical** and **numeric**. We will investigate what variables are in more detail in the next chapter.

Let's go through a few types of polls and surveys:

1   *Telephone polls and surveys* have been used historically, as well as *calling door to door* asking people to fill them in. Pre-Covid, door-to-door requests still resulted in a high response rate; however, the high costs associated with hiring the personnel required and the time taken to get the data in the correct format continued to be challenging obstacles to overcome for many organisations and data collection agencies.

2   *Popular online survey platforms* include Qualtrics, Slido and SurveyMonkey. Like many other survey platforms, SurveyMonkey and Slido are free, but others, like

Qualtrics, require payment. There are usually options to download any data generated into Excel files or other formats to enable further data analysis and visualisation. There are also in-built functions to do these things within platforms like Qualtrics, making it one of the more popular platforms to use for professionals.

SurveyMonkey has created an extremely useful webpage which outlines simple steps to create an online survey, which is something you may wish to consult if you are thinking of creating your own survey for an associated research project: https://www.surveymonkey. co.uk/mp/online-polls/. Emphasis is placed on knowing your audience, which can help to dictate the language used and set up of the overall appearance of the survey.

A great way to learn about creating a good poll or survey is to look at what others have done, and there are so many to look at and even participate in! In addition, many media articles and outlets use information from them to create interesting and engaging data stories.

Some questions in surveys and polls ask users to tick a response from a set of options. Responses are often measured using categorical variables (like favourite flavour of dessert, or grade obtained in school). Such questions can include the use of a Likert scale. These scales are commonly used to measure attitudes, knowledge, perceptions, values, and behavioural changes. A Likert-type scale involves a set of options that respondents may choose from to rate their responses to evaluative questions. A common example on a five-point Likert scale is 'strongly disagree', 'disagree', 'neither agree not disagree', 'agree', 'strongly agree'.

Surveys and polls can also include open-ended questions, which require users to respond using written sentences, to measure attitudes, knowledge, perceptions, values, and behavioural changes. Many surveys, especially workplace examples, tend to include an open-ended question that asks the user for any other comments, or any other relevant information.

· · · · · · Develop your skills! 3.1 ·

There is a good chance that you have already filled in numerous online surveys on a variety of topics. For this 'Develop your skills' task, search for at least five different online surveys on topics of your choice. Feel free to fill them in and see if the survey presents any data after you have submitted your responses. When looking through these different online surveys, make notes based on the following prompts:

1 Why did you choose the theme/topic of the survey you searched for?
2 Comment on the length of the survey (i.e., is it too long or short or the right length?).
3 Is the survey visually appealing? Explain your answer.
4 Did the website provide you with any data after you submitted your results? If it did, did you understand the data display? What did it show?

· ·

3.2 Language used in polls and surveys, and improving response rates

Figure 3.1 Language
Source: Photo by Clarissa Watson on Unsplash

The intended audience of a poll or survey will have a significant bearing on the language used in it. The main aim of the poll or survey is to have a representative sample, so that the findings can be generalised and/or extended to groups and other individuals who are like those included in the sample. Using appropriate language that grabs the attention of the intended audience can help to increase completion rates from the intended audience. Qualtrics have produced an online resource providing guidance on how to increase response rates in online surveys: https://www.qualtrics.com/uk/experience-management/research/improve-survey-response. Many of the suggestions in there can also be applied to online polls.

The information provided in the Qualtrics resource discusses several extremely relevant points that can help to increase response rates in polls and surveys. These include:

1   Using incentives for the participants (such as a coupon or discount for other products).
2   Using cognitive dissonance (which can include appealing to the participants' values and beliefs to convince them to take part).
3   Timing of asking for the feedback (asking immediately after a purchase, for example, usually results in higher response rates).
4   Keeping them short and focused (especially for polls).
5   Sending gentle reminders to participants.

Try to avoid:

1   Assuming polls and surveys are representative of the populations the samples are taken from.
2   Assuming authors of polls and surveys factor in the importance of clear and understandable language for their target audience.

·····Develop your skills! 3.2 ···

For this exercise you will need to use your notes taken for the previous one. Look at the notes you made about the online surveys you searched. Choose two of the online surveys and make additional notes.

Comment on the language of each survey. Is it clear and concise? Written well? (Think about the guidelines from Chapter 2, 'Building a Foundation'.) Are there any words in there that you didn't understand, even after trying to figure out what they mean? If it makes it easier, arrange your notes into columns. You could have column headings titled 'Good' and 'Bad', or 'Happy' and 'Not Happy', for example.

·····Develop your skills! 3.3 ···

In the following set of questions, you will continue to develop your skills in being able to critique data displays, as well as looking at the key features of an online survey based on a skills assessment.

Figure 3.2   Skills assessment
Source: Photo by Cookie the Pom on Unsplash

For this exercise, you will need to use the website titled 'National Careers Service: Skills assessment' (https://nationalcareers.service.gov.uk/skills-assessment). Scroll down and click on the pink button labelled 'Go to discover your skills and careers'. Then scroll down and click on the pink button 'Start assessment'. Fill in the survey, and answer the following questions:

1   Comment on the language of each survey. Is it clear and concise? Written well? (Think about the guidelines from Chapter 2, 'Building a Foundation'). Are there any words in there that you didn't understand, even after trying to figure out what they mean? If it makes it easier, arrange your notes into columns. You could have column headings titled 'Good' and 'Bad', or 'Happy' and 'Not Happy', for example.
2   Are there any features that stand out in the survey? Anything that is useful for the user, to help them fill in the survey?
3   Who do you think are the target audience of this survey?

·····Develop your skills! 3.4 ··········

In the following set of questions, you will continue to develop your skills in being able to evaluate data displays, as well as looking at the key features of an online survey based on an assessment of perfume preferences.

Figure 3.3 Perfume
Source: Photo by William Boyd on Unsplash

For this exercise you will need to use the Penhaligon's website (https://www.penhaligons.com/uk/en/fragrance-profiling).[1] Click on the pink button with text inside labelled 'Shall we begin'. Fill in the survey, and answer the following questions, related to this webpage:

1   Comment on the language of each survey. Is it clear and concise? Written well? (Think about the guidelines from Chapter 2, 'Building a Foundation'). Are there any words in there that you didn't understand, even after trying to figure out what they mean? If it makes it easier, arrange your notes into columns. You could have column headings titled 'Good' and 'Bad', or 'Happy' and 'Not Happy', for example.
2   Are there any features that stand out in the survey? Anything that is useful for the user, to help them fill in the survey?
3   Who do you think are the target audience of this survey?

· ·

---

[1]This example is not an endorsement of Penhaligon's. It was selected merely as an example of the language used in an online survey.

3.3 Sampling

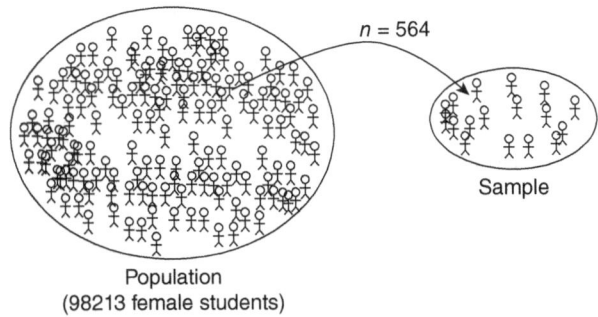

$n = 564$

Sample

Population
(98213 female students)

Figure 3.4 Taking a sample from a population

Collecting data using a survey or poll, from a **sample** that has come from a population of interest, is often faster, cheaper and more practical than trying to get data from the whole population. Figure 3.4 demonstrates taking a sample from a population of interest, in this case female students. Poll and survey reports should include the target population, sampling method, sample size, date and the exact questions asked.

Random sampling helps to avoid subjectivity in choosing participants and allows for the calculation of **sampling error**. The larger the sample taken, the better, in terms of it being representative of the population of interest.

Sampling is a crucial aspect of research and data collection, allowing us to make inferences about populations based on a smaller subset of individuals. Various types of sampling methods exist, each with its own advantages and limitations:

1   **Random sampling**, also known as probability sampling, involves selecting individuals from a population in a completely unbiased manner, giving everyone an equal chance of being chosen.
2   **Stratified sampling** involves dividing the population into distinct subgroups, or strata, and then selecting individuals from each stratum in proportion to their representation in the population. This method ensures that each subgroup is adequately represented in the sample, making it useful when studying heterogeneous (i.e. mixed) populations.
3   **Systematic sampling** involves selecting individuals at regular intervals from a population list (e.g. for every 50th individual), which can be efficient and less time-consuming.
4   **Cluster sampling** entails dividing the population into clusters, such as geographic regions or schools, and randomly selecting entire clusters to include in the sample. This method is useful when it is impractical or costly to sample individuals directly.

5   **Convenience sampling** involves selecting individuals who are readily available
    and willing to participate, making it a convenient but potentially biased sampling
    method.

Understanding these different types of sampling enables researchers to make informed
decisions about the appropriate approach to use in their studies, considering the objec-
tives, resources and constraints at hand.
    Try to avoid:

1   Confusing or mixing up data taken from a sample, with data taken from a whole
    population.

······ Develop your skills! 3.5 ·· ···

For the following statements, decide whether they are true or false and circle the correct answer:

1   Taking a smaller sample is often better than taking a bigger sample from a
    population of interest.

    True or False?

2   Taking a sample from a population is often cheaper and faster than trying to obtain data
    from the whole population.

    True or False?

3   Random sampling helps to avoid subjectivity in choosing participants.

    True or False?

3.4 Sampling and non-sampling errors

Sampling errors occur because of taking a sample, affecting how representative it is
of the population it came from. Sampling errors have the potential to be bigger in
smaller sample sizes.
    Non-sampling errors, which include errors such a selection bias and inter-
viewer effects, can be much larger than sampling errors and are always present. They
are often impossible to correct for after the poll or survey has taken place. Any
potential non-sampling errors must be minimised in the design phase of the poll or
survey.

Bias occurs when the data collected are consistently over- or underestimated. This could lead to results that are inaccurate. A **biased selection** process is an inadequate selection or sampling process, types of non-sampling errors that might lead to this include the following:

Selection bias. The population sampled is not exactly the population of interest. For example, asking the readers of *Vogue* magazine (USA) in an online survey for their opinions on same-sex marriage, then generalising the results to all US people.

Non-response bias. Choosing a certain group of people to be surveyed but they fail to respond, for example because they refuse to participate or are unreachable.

Self-selection bias. This is where people choose to volunteer to take part in a poll or survey, rather than being randomly selected. For example, a TV show could present a poll question and ask viewers to phone in or to respond online. The viewers themselves decide to participate.

Question effects. Variations in wording can influence responses. For example, compare 'Do you think there is a lack of discipline in children? Do you believe in national conscription?' and 'Do you think children should be given the time and space to grow, and play with other children? Do you believe in national conscription?'

Interviewer effects. Different interviewers asking the same question can obtain different results. This may be down to the sex, race, or religion of the interviewer. For example, when asking male participants 'How often do you feel depressed?', a female interviewer found that 21% answered 'Often', while a male interviewer found 6% gave that answer.

Behavioural considerations. People tend to answer questions in a way they consider socially desirable. For example, pregnant women being asked about their drinking habits are unlikely to admit to consuming alcohol, and people are more likely to say they have never cheated on a partner, even when they have.

Transferring findings. Taking the data from one population and transferring the results to another. For example, Londoners' opinions may not be a good indication of the opinions of people from Edinburgh.

Survey-format effects. These can occur when the layout or order of questions is presented differently to different people or groups. Examples are question order, survey layout, and whether interviewed by phone, in person or by mail.

Try to avoid:

1   Mixing up selection and self-selection bias. They look similar; however, they have
     different meanings.
2   Becoming too cynical when recognising that a lot of data collected using polls
     and surveys is susceptible to all sorts of non-sampling biases. Even though these
     can be present, the data collected from polls and surveys are still extremely useful.

··· ··· Develop your skills! 3.6 ··· ··

Figure 3.5   Netflix sign on a building at sunset, Hollywood, Los Angeles, 2021
Source. Photo by Venti Views on Unsplash

Read the following scenario and answer the questions that follow:

> A link to an online survey was emailed to Netflix customers, asking them several
> questions, including 'Roughly how many hours do you spend a week watching Netflix?'
> and 'How many people in your household use Netflix?' Netflix then use the results to
> report on their users in their monthly bulletin to service managers.

1   Which non-sampling errors are present in this survey?
2   Are there any issues with the questions presented in the scenario?

For the following statements, decide whether they are true or false and circle the correct answer:

3   Non-sampling errors are often bigger than the random sampling errors in surveys.

     True or False?

4   People will sometimes answer a question differently for different interviewers.

     True or False?

5   Slight changes in the wording of questions can often make a big change to survey results.

    True or False?

6   Non-response will not cause bias in surveys because non-respondents are likely to behave the same (and fill in the survey in the same way) as the people who responded.

    True or False?

· ·

· · · · · **Develop your skills! 3.7** ·

Figure 3.6   Amazon delivery box with gifts in front of a Christmas tree.
Source: Photo by Wicked Monday on Unsplash

Read the following scenario and answer the questions that follow:

A link to an online survey was emailed to customers who shop on Amazon online, asking them several questions, including 'Roughly how often do you shop on Amazon?' and 'How many people in your household use Amazon?' Amazon then use the results in their annual report, which includes data to help determine the budget for their online marketing content.

1   Which non-sampling errors are present in this survey?
2   Are there any issues with the questions presented in the scenario?

For the following statements, decide whether they are true or false and circle the correct answer:

3   Non-sampling errors are often smaller than the random sampling errors in surveys.

    True or False?

4   People will always answer a question differently for different interviewers.

    True or False?

(Continued)

5   Slight changes in the wording of questions can often make a small change to survey results.

True or False?

6   Non-response will cause bias in surveys because non-respondents are unlikely to behave the same (and fill in the survey in the same way) as the people who responded.

True or False?

· ·

· · · · · · Develop your skills! 3.8 ·

Figure 3.7   Petrol station in Bicester, UKS
Source: Photo by Kyle Bushnell on Unsplash

Read the following scenario and answer the questions that follow:

A survey was administered at a series of petrol station stores across the North of England, asking people who filled up with fuel several questions, including 'How many hours do you spend in a petrol station each month?' and 'Are you a member of the rewards scheme offered at the petrol station?'

1   Which non-sampling errors are present in this survey?
2   Are there any issues with the questions presented in the scenario?

· ·

Key points to remember

1   Make sure you do not confuse sampling and non-sampling errors. Sampling errors occur because of taking a sample and are measurable. Non-sampling errors (which can include errors such a selection bias and interviewer effects) are usually unavoidable and very difficult to correct for after the data have been collected.

2   Sampling errors occur because of taking a sample and are measurable.
3   Non-sampling errors are usually unavoidable and very difficult to correct for after the data have
     been collected.
4   Polls and surveys are not always representative of the populations the samples are taken from.
5   Don't assume authors of polls and surveys factor in the importance of clear and understandable
     language for their target audience.

References to support this chapter

National Careers Service (2022) Skills assessment. https://nationalcareers.service.gov.uk/
     skills-assessment (accessed 12 August 2022).
Penhaligon's (2022) Fragrance profiling. https://www.penhaligons.com/uk/en/fragrance-profiling
     (accessed 12 August 2022).
Qualtrics (2021) How to increase survey response rates. https://www.qualtrics.com/uk/
     experience-management/research/improve-survey-response/ (accessed 11 March 2021).
SurveyMonkey (2021) How to create a poll in 3 simple steps. https://www.surveymonkey.co.uk/mp/
     online-polls/ (accessed 11 March 2021).

4

INTRODUCTION TO CATEGORISING AND VISUALISING DATA

An overview of this chapter

Now that we have covered some foundational knowledge on sampling, and the importance of literacy skills, we will move onto key concepts in statistics. This chapter will dispel the myth that statistics is the same as mathematics. It will include a series of examples demonstrating how data are arranged and visualised, as well as ensuring you appreciate the need to dig beneath the numbers and data displayed, to gain an appreciation for the context or subject. Without this, the numbers become meaningless.

Real-world data sets will be used, with support and guidance provided, to demonstrate the many ways that data can be displayed. This chapter will focus on the different types of variables that exist (a variable can be defined as a characteristic that can be measured and that can have different values; height, age, income, province or country of birth, grades obtained at school and type of housing are all examples of variables), along with the ways in which they can be displayed and used in data stories. Through guided tasks, you will begin to recognise the many shades of grey that exist within statistics. This will then link well with the next chapter, which looks at some of the more specialised terms used in statistics.

Resources to support this chapter

This chapter draws on a range of sources to help introduce you to the foundations of statistics. In this chapter, we use a range of real-world examples to show you how data are often structured and reported, as well as how we measure uncertainty. You can build your skills by trying the suggested activities which will help you to build your confidence in being able to identify different types of variables, as well as the data that sit within them. Activities will also encourage you to explore different data displays, and their value in communicating data trends.

Resource	Date accessed	Location
FBI bank crime statistics, *FBI*, 2019	23 August 2021	https://www.fbi.gov/file-repository/bank-crime-statistics-2019.pdf/view
Number of banks in the USA, 2019	23 August 2021	https://www.google.co.uk/search?q=number+of+banks+in+usa&ie=UTF-8&oe=UTF-8&hl=en-gb&client=safari
List of largest banks in the United States, *Wikipedia*, 2021	23 August 2021	https://en.wikipedia.org/wiki/List_of_largest_banks_in_the_United_States

The websites referred to in all the activities are provided as part of this book's online resources. You can find them at **https://study.sagepub.com/jonesstatsliteracy**.

4.1 Variables, levels of measurement and descriptive data

Being able to identify the types of variables you are working with is an important first step in being able to decide the best way to display, report and communicate data. To better help you understand what a variable is, it can be defined as a characteristic that can be measured and that can have different values. Height, age, income, province or country of birth, grades obtained at school and type of housing are all examples of variables. If we were looking at a variable like grades obtained, an A or B grade would be examples of data for that variable. Variables may be classified into two main categories: categorical and numeric. Figures 4.1–4.3 display the different types of variables, with examples, that are often measured and recorded.

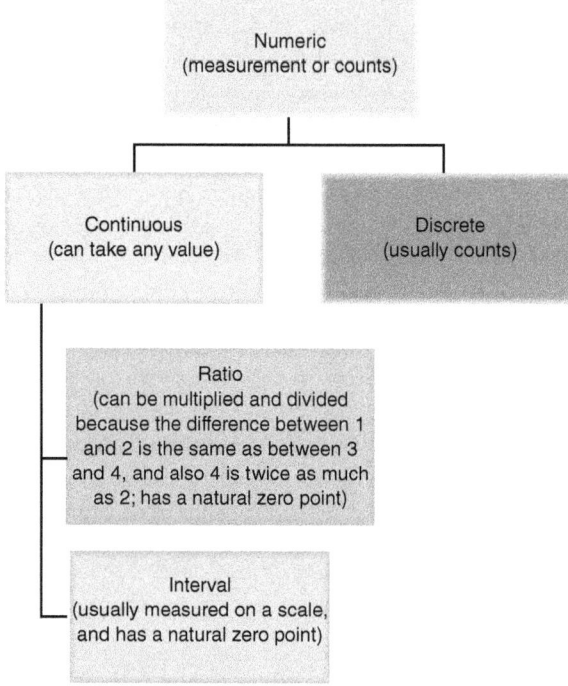

Figure 4.1   Types of numeric variables

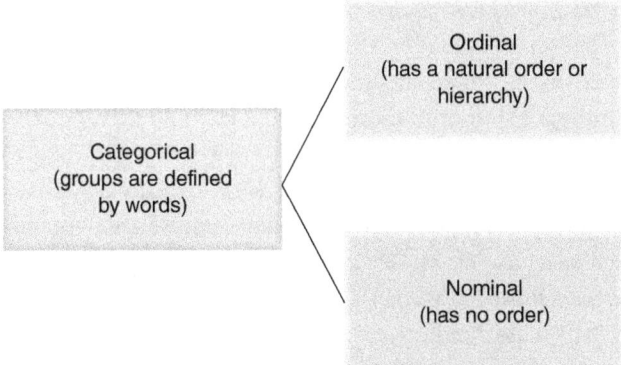

Figure 4.2 Types of categorical variables

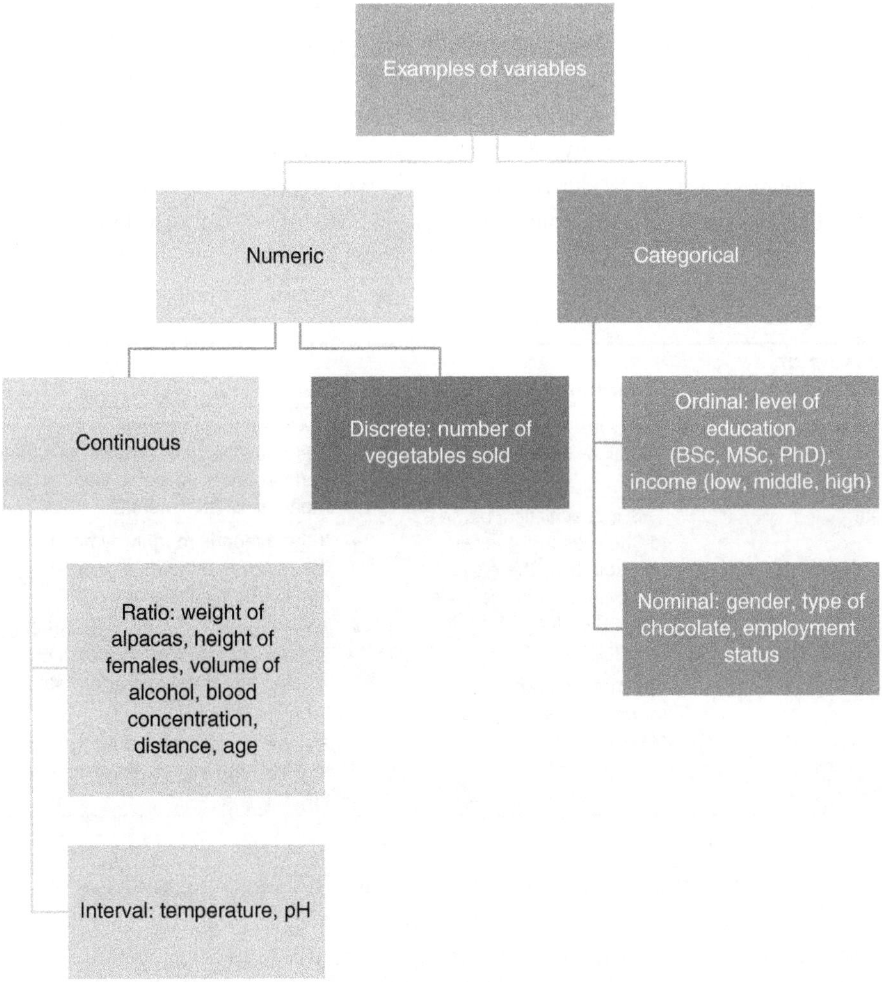

Figure 4.3 Examples of variables

Some of the examples given above can fall into more than one group. For example, the type of occupation measured in a group of people (listed as nominal in Figure 4.3) could also have an order or hierarchy (i.e., shop floor assistant, assistant manager, manager, in order from most junior to most senior) and would then be an ordinal variable. In addition, some of the variables mentioned can be arranged differently to become other types of variables. For example age, which is normally a **ratio** type of variable, can become categorical, and specifically ordinal, if it is arranged into groups that might need to be ticked in a box: 15–21, 22–25, 26–30, etc.

Now that we have looked at variables and what they are, we will move onto **descriptive statistics**, which are often used to summarise a given data set, which can be either a representation of the entire population or a sample of a population. Descriptive statistics are broken down into measures of central tendency and measures of variability (spread). Central tendency is a fancy way of saying where the middle of the data sits. Table 4.1 lists three ways to measure it. The art of using descriptive data (i.e., describing what you see in the data, which could include reporting average and standard deviation values) is often overlooked, but it can be used to craft a beautifully coherent data story (Chapter 10). Using descriptive data can include the use of frequencies of a variable of interest, (e.g., number of tattoos). It can also include a description of central values and how spread out the data are. The main advantage of using descriptive data is that most people often understand what they are, being able to interpret simple graphs and tables that display this information (Bach et al., 2016, 2018). Explained well, descriptive data can bring a story to life, producing a more compelling point of view or set of conclusions.

Table 4.1 Central tendency

Measure of central tendency	Definition	Notes
Median	This is the middle value when the values are arranged in order	• If we had to summarise all the observations on a variable as a single number, then we would want to use a measure of their 'centre'
Mean	This is a statistical name for the ordinary, everyday average. It is where the dot plot balances. Sometimes also called the arithmetic mean.	• If the shape is roughly symmetric, the mean and the median will be approximately the same
Mode	The most frequently occurring data point in the set of observations	• If the shape is strongly skewed there is no single compelling notion of 'centre', and the mean and median can be quite different
		• Means can be skewed by outliers in smaller data sets whereas medians are not

Descriptive statistics can sometimes give us an idea of the central tendency of a data set. Central tendency provides important information as to where the data points sit. If data are presented in a graph, it helps determine what the data spread looks like and whether the data look symmetrical. The two most used measures of centre are the

median (in a box-and-whisker plot this is denoted by a vertical line) and the **mean** (which is sometimes given denoted by a triangle symbol) (Figure 4.4). The median is the middle value of a set of data points, whereas the mean is a statistical name for the ordinary, everyday average. Another measure of centre is called the **mode,** which is the most frequent data point present in the set of observations.

The mean is a fundamental measure of central tendency used to summarise a set of data values. It is commonly referred to as the 'average' and is calculated by adding up all the values in a data set and dividing the sum by the total number of values. The mean provides a representative value that balances out the individual observations, giving us a sense of the typical value in the data set. It is widely used in various fields, such as statistics, mathematics and social sciences, to analyse and interpret data. The mean is particularly useful when dealing with continuous or interval variables, as it considers all the observations and provides a single value that can be easily compared across different data sets. However, it is worth noting that extreme outliers can significantly influence the mean, making it sensitive to extreme values. Therefore, it is important to consider other measures of central tendency, such as the median or mode, in conjunction with the mean to obtain a comprehensive understanding of the data and variables being investigated.

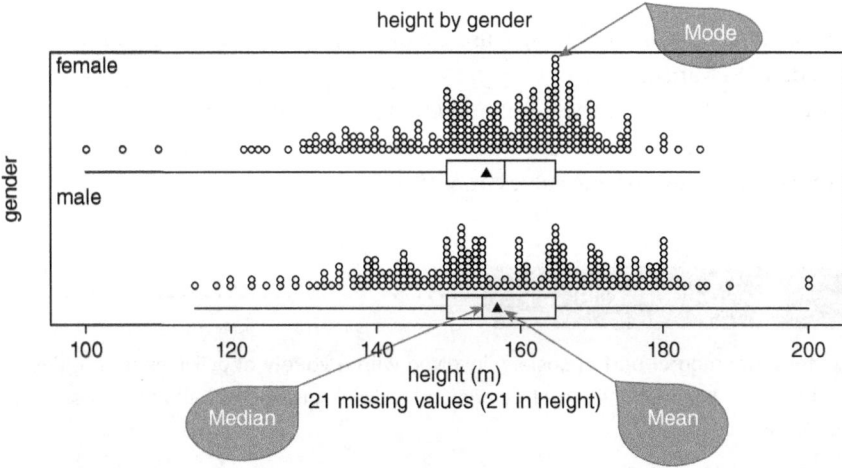

Figure 4.4   Central tendency and measures of centre and spread

But the central tendency is not the whole story. We are often interested in how spread out the data are (Table 4.2). If we put the data into a graph, what does the data spread look like? Is it symmetrical? Have a look at the Section 4.2 on visualisations and ask yourself these questions when looking at dot plots.

The **standard deviation** of a set of data can be thought of as the average or typical distance between the data points and the mean, and this gives the reader an idea as to how spread out the data are.

Table 4.2   Spread

Measure of spread	Definition	Notes
Standard deviation	This is the square root of the average of the squared distances between the data points and their mean (average the squared distances and then take the square root of the answer). It can be thought of as 'the average of the distances between the data points and the mean'.	When a dot plot is a roughly symmetric mountain shape then typically about 68% of the data points fall within one standard deviation either side of the mean
Interquartile range (IQR)	This is the difference between the upper and lower quartiles of the data points. You can think of the lower quartile as the median of the lower half of the values arranged in order, and the upper quartile as the median of the higher half of the values arranged in order (see Figure 4.9, p. 43).	A smaller (larger) standard deviation means the observations are less (more) different from one another

Calculating a mean and standard deviation for a sample or population can be done by hand. However, there are many online calculators that will do this for you – for example, https://www.calculator.net/standard-deviation-calculator.html. There are also many software platforms that will also do this for you, iNZight for example.

Try to avoid:

1   Describing data as continuous, discrete or categorical. These terms of usually used to describe variables.
2   Automatically stating the mean and always using this to measure the centre of your data. Sometimes the median is a better presentation, especially if the sample of data includes extreme values or outliers.

······· Develop your skills! 4.1 ·· ······································

Banks form an integral part of society, involved with a variety of activities on a global scale. Sadly, they are also often the target for criminal activity. Answer the following questions based on this topic:

1   Do a Google search for the number of banks in the USA. How many are there? Does this number surprise you?
2   Check out this website, which looks at the numbers of bank robberies in the USA in 2019: https://www.fbi.gov/file-repository/bank-crime-statistics-2019.pdf/view. Which type of bank had the highest number of robberies in 2019?
3   Data from Wikipedia on the largest banks in the USA have been transferred into an Excel spreadsheet for you to use, which you can find as part of the online resources (https://study. sagepub.com/jonesstatsliteracy). Once you've opened up the spreadsheet, take a look at column B. What type of variable is 'Bank name'?
4   Go to the following webpage: https://lite.docker.stat.auckland.ac.nz/. Click on File then Import Dataset. Load the Excel spreadsheet used in Question 3. Click on Variable selection and Select

first variable, selecting: Total assets (billions of US$). Then click on Summary and state the mean and standard deviation (SD) from this sample of the largest banks in the USA.

5   Using the same variable in Question 4, state the median.
6   Explain why the median and mean values stated in Questions 4 and 5 are different. Look at the plot to help you answer this question.

. . . .

4.2 Visualising data

Tables and graphs (also sometimes called plots or charts) are very common ways of presenting data, and you have probably come across them before, in a variety of displays and outlets. They are found in many disciplines, as well as being present in everyday life, from media reports to social media platforms. In this section, you will learn how to interpret graphs, and we will cover the main types used in statistics. Being able to produce graphs from data usually enables the user to spot features in the data much more easily. This section will help you to hone your skills in these areas, as well as asking useful questions to interrogate data displays.

Figure 4.5   Example of visualising data
Source: Photo by Morgan Housel on Unsplash

Whenever you are looking at a graph (as a recap, remember the horizontal axis is usually called the x-axis, and the vertical axis is usually called the y-axis) you need to ask yourself these questions:

1   What are the main features I can see in the graph?
2   What are the main features of the graph telling me?
3   What other details are useful for understanding the variable or answering any questions that I may have?
4   What other questions do I have?

You may not be able to answer all these questions yet; however, they are useful ways to help you to think about data critically. We will revisit these questions throughout this chapter and elsewhere in the book.

The type of graph used to display data will depend on what type of data you have. For numeric data, useful graphs to display this information can include histograms, scatter plots, dot plots and box-and-whisker plots. If you have categorical data, bar graphs and side-by-side plots are often useful.

The best type of graph to illustrate the relevant data will also depend on the numbers of variables you are comparing, as well as the number of groups. Table 4.3 provides a useful overview of the strengths and weaknesses of commonly used graphs.

Table 4.3 Strengths and weaknesses of commonly used graphs

Type	Strengths and weaknesses	Examples
Dot plot	• Retains numerical information, can also check for skewness of data, modality, and outliers	Gender and salary
Scatter plot	• Can give you useful information on whether the axes in the graph are associated with each other • Retains numerical information	Age and salary
Box-and-whisker plot	• Very good for comparing several data sets • Displays centre and spread of data • Not useful for small data sets	Gender and salary
Histogram	• Displays relative density of observations (i.e., gives a good idea of the shape of the distribution) • Good for large amounts of data	Heights
Bar chart	• Can be used to present a variety of data, easily communicated (e.g., percentages)	Gender and race

Some variables (numeric or categorical) can involve data collection over time, which we call time series data. This is especially relevant for disciplines that investigate changes over time, such as the sciences, social sciences, psychology, history and public health. Time series can include changes over short or long periods.

Try to avoid:

1   Giving short and limited descriptions of plots.
2   Selecting the incorrect type of graph to display data (e.g., using a histogram to display a categorical variable).

······· Develop your skills! 4.2 ··· ·······································

Using the data in the Excel spreadsheet 'Banks in USA 2020', which you can find as part of the online resources (https://study.sagepub.com/jonesstatsliteracy), answer the following questions:

1   Which type of plot would be useful to explore the variable 'Headquarters location'?
2   Which type of plot would be useful to explore the variable 'Market capitalisation (billions of US$)'?
···

4.2.1 Histograms

Histograms are an extremely useful and common way of representing data (such as lengths, widths, and heights). They are like bar charts but show the frequency density instead of the frequency. They can also be used to determine information about the distribution of data.

Figure 4.7 is a histogram of the mandible length of a sample of dolphins. Check out the boxes in Figure 4.7 to give you an overview of what histograms display.

Figure 4.6 Image of a dolphin

Source: Photo by Fabrizio Frigeni on Unsplash

· · · · · ·Develop your skills! 4.3 ·

1   What are the advantages of using a histogram to display data?
· ·

4.2.2 Dot, box, and scatter plots

A **dot plot**, also known as a strip plot or dot chart, is a simple form of data visualisation that consists of data points plotted as dots on a graph with an *x*- and *y*-axis. These types of charts are used to graphically depict certain data trends or groupings. They can be expanded to include **side-by-side plots**, for example when you are looking at a ratio variable (e.g., weight) subdivided by a categorical viable (e.g., gender).

These types of plots are very useful for showing distributions of data, which can help you to identify any interesting patterns present. The example used in this section looks at parks in New Zealand. Parks are great places to go for a stroll, whether you are out with the family and pets, or with friends or even on your own. They come in all shapes and sizes, and can include things like lakes, hills, and coffee shops!

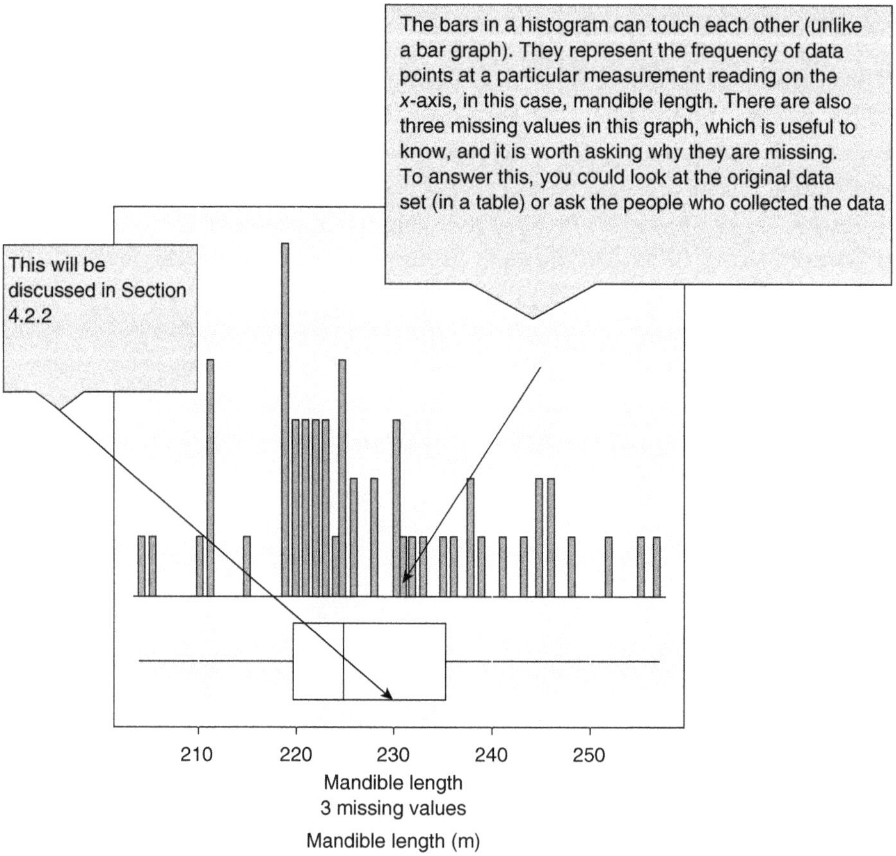
The bars in a histogram can touch each other (unlike a bar graph). They represent the frequency of data points at a particular measurement reading on the x-axis, in this case, mandible length. There are also three missing values in this graph, which is useful to know, and it is worth asking why they are missing. To answer this, you could look at the original data set (in a table) or ask the people who collected the data
This will be discussed in Section 4.2.2
210
220
230
240
250
Mandible length
3 missing values
Mandible length (m)

Figure 4.7 Histogram and mandible length

Figure 4.8 Image of parks

Source: Photo by John McCann on Unsplash

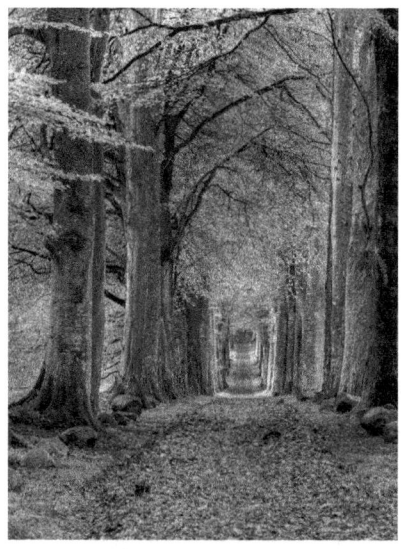

Figure 4.9 shows the lengths of a series of parks in New Zealand. Check out the boxes associated with these plots. This type of display is very useful since it includes the numerical information (Park length), and we can also check for skewness of data, modality, and outliers, from the dot plot. The **box plot** is also very useful since it shows the centre and spread of data. Check out the boxes associated with these plots, to give you more information about dot and box-and-whisker plots.

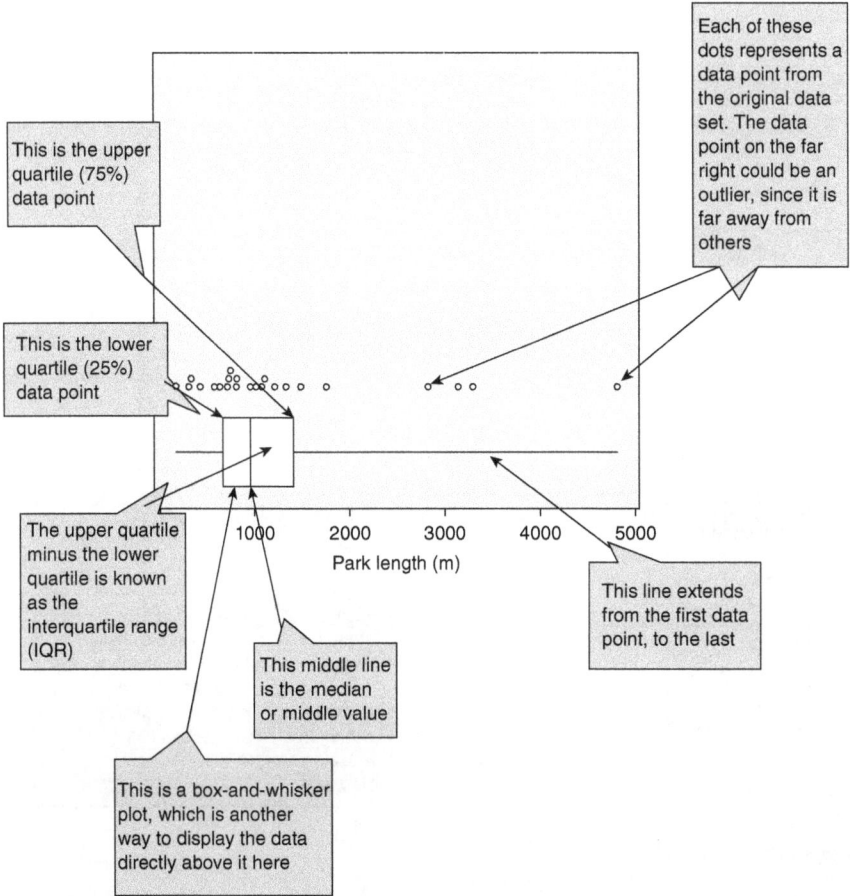

This is the upper quartile (75%) data point

Each of these dots represents a data point from the original data set. The data point on the far right could be an outlier, since it is far away from others

This is the lower quartile (25%) data point

The upper quartile minus the lower quartile is known as the interquartile range (IQR)

This middle line is the median or middle value

This line extends from the first data point, to the last

This is a box-and-whisker plot, which is another way to display the data directly above it here

Park length (m)

1000   2000   3000   4000   5000

Figure 4.9   Dot plots and box-and-whisker plots

The **scatter plots** in Figures 4.10 and 4.11 present data based on the length and area of a sample of parks in New Zealand. These figures are extremely useful, in terms of the displays selected, since they can give you useful information on whether the variables in the graph are associated with each other, and they also retain the numerical information about the variables. Figure 4.10 shows how all the data points are dispersed, how far away they are from each other, and they also show whether there are any outliers (i.e., data points further away from others). Figure 4.11 has a trend line over imposed on the data points to show the two variables in the plot are associated with each other. If more of the data points fall close to the line, you can say there is a strong association. When you progress onto more advanced types of statistical tests (i.e., correlation coefficients – this is a numerical measure of a type of association), you can obtain numerical values to help you ascertain how strong this association is between variables in a scatter plot. Have a look at the boxes to find out more!

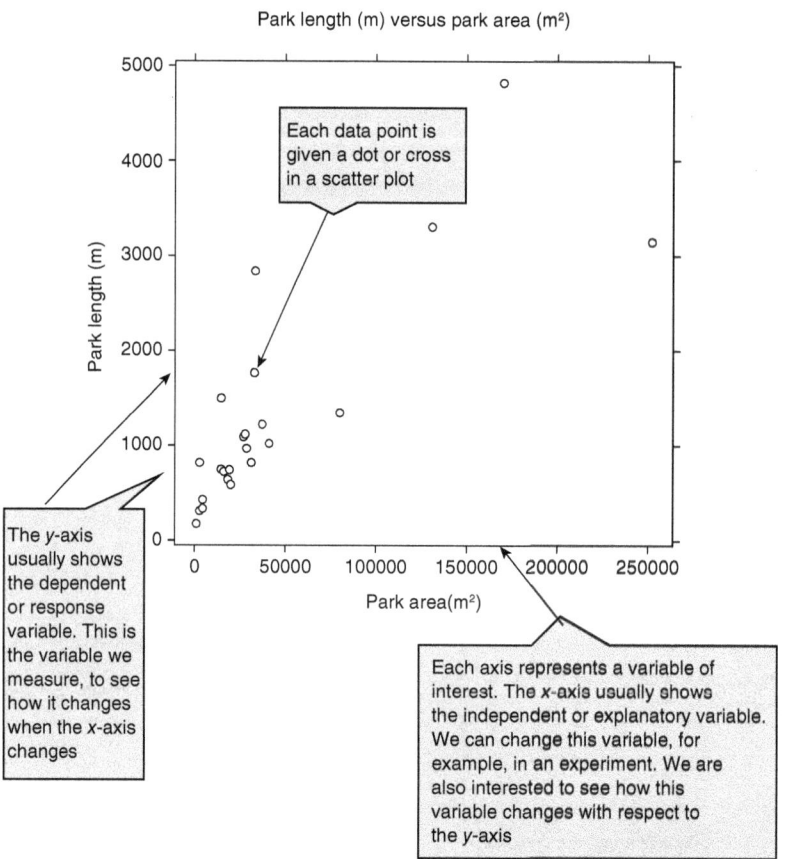

Figure 4.10 Scatter plot

When looking at scatter plots, we need to think about four things:

Trend. Is there a linear (straight-line) relationship – in other words, can you roughly fit all the dots onto a straight line? Or do the dots appear to fit a curved line? In the latter case we could say there appears to be a nonlinear relationship. Alternatively, if the dots do not seem to show any pattern, we could say there is no relationship.

Scatter. If you were to draw a straight line though the data points, are they close to the line? Alternatively, is there a lot of variation in data points around the line? For example, look at the scatter plot with the line drawn on top of the data points (Figure 4.11). If the scatter of the data were more evenly spread in this graph, we would expect the data points to be closer to the line, and not so spread out.

Strength of relationship. If the data points are close to a line we draw as closely as possible to all the data points, then we can say there is a strong relationship between the variables plotted. If the data points are spread out and far away from each other, with no discernible pattern, then we say there is a weak relationship.

Association. If the data points on the y-axis increase in line with the x-axis, as is the case in Figure 4.11, then we can say there is a positive association. If, however, the data points decrease on the y-axis as the x-axis increases, then we say there is a negative association.

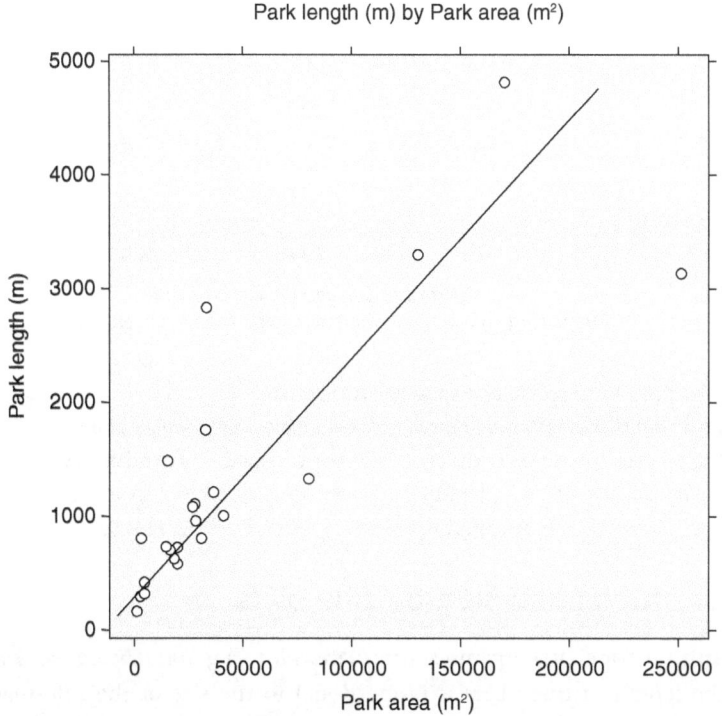

Figure 4.11   Scatter plot with trend line

Try to avoid:

1   Automatically assuming extreme data point values are outliers. Context-specific information is needed to ask if an extreme data point really is an outlier. This can be gained from subject experts, or potentially from the people who collected the data.
2   Removing perceived outliers from plots.
3   Explaining relationships in scatter plots in terms of one variable causing another variable to increase or decrease. Causal relationships can be ascertained, and will be more accurate from using inferential techniques, not just by visually describing the relationship observed in a scatter plot.

····· Develop your skills! 4.4 ····· ···································

Using the data in the Excel spreadsheet 'Banks in USA 2020', which you can find as part of the online resources (https://study.sagepub.com/jonesstatsliteracy), answer the following questions.

The following scatter plot was generated using iNZight, to explore the relationship between the variables 'Total assets (billions of US$)' and 'Market capitalisation (billions of US$)'.

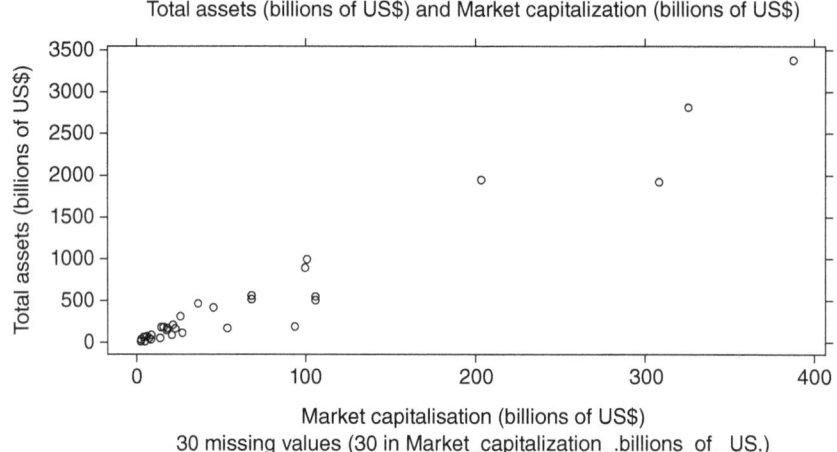

Total assets (billions of US$) and Market capitalization (billions of US$)

Market capitalisation (billions of US$)

30 missing values (30 in Market_capitalization_.billions_of _US.)

1   Do the variables in the scatter plot appear to be related?
2   Describe, in detail, the relationship between the variables in the scatter plot.
3   Reflecting on your answers in Question 2, have you changed your mind as to whether you think there is a relationship between the variables in the scatter plot? Explain your answer.

···

4.2.3 Bar graphs and side-by-side plots

A **bar graph** (or bar chart) organises information into a graphic using bars of different lengths. The length of these bars is proportional to the size of the information they represent. For example, a vertical bar graph can show the popularity of different colours among a group of children, or the main car of choice of countries in the European Union. Bar graphs are extremely useful for displaying data from a table of counts.

Side-by-side plots are extremely useful ways of splitting a ratio variable by a categorical variable. It can help you to further explore any patterns you observe in the data. For example, a ratio variable like Weight or Salary might be bimodal; when it is subdivided by a categorical variable like Sex, it can help to explain the bimodal split (since on average, males tend to be heavier than females). Statistical data often include variables like Sex and Gender, and many often only include males and female as potential responses or forget to include gender-diverse examples. The Sex of an individual often refers to the sexual appearance of how an individual is born, whereas the Gender of an individual is how they self-identify (e.g., a person born as a male may self-identify as a female, and she would say her Gender is female). This will have serious implications for how the surveys are used, as well as how accurately they represent the views of cross-sections of society.

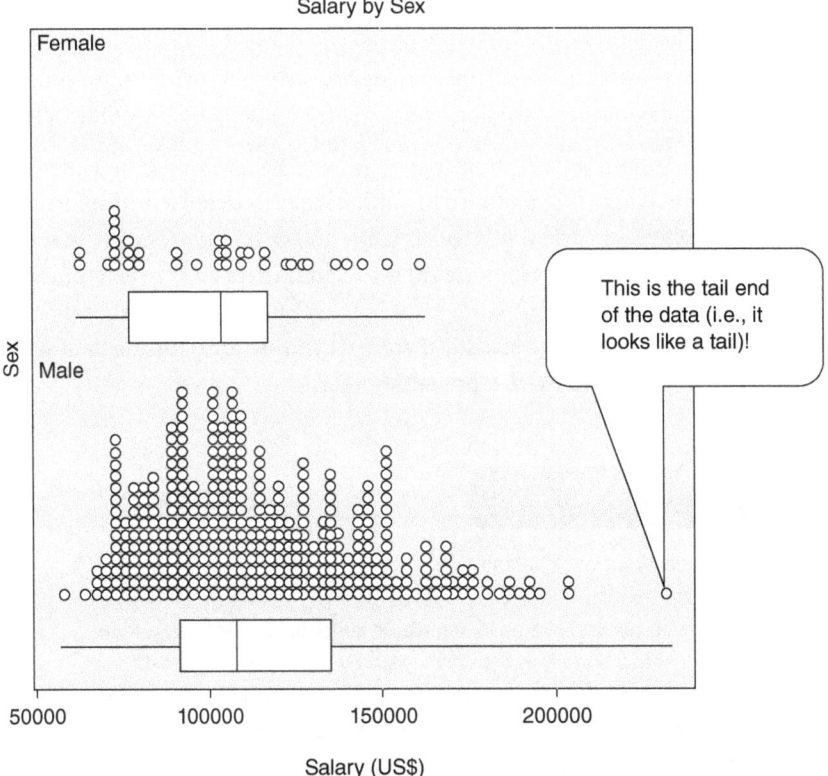

Figure 4.12 Side-by-side dot and box-and-whisker plots (on the same scale)

Questions you should ask yourself when looking at graphs will depend on the type of graph you are looking at, to be able to spot a particular feature about the data. The following features of a graph are a useful starting point:

Skewness. When looking at dot plots or histograms, how the data are spread out will tell you if they are positively skewed, negatively skewed, or symmetrical. For example, if we look at Figure 4.12, which shows Salary by Sex (side-by-side plots) the tail end of the male data is located towards the positive *x*-axis values. We therefore call this positive skew. If the tail end of the male data points were located towards the increasingly negative *x*-axis part of the graph, we would call this negative skew.

Symmetry. Assessing the symmetry of your data points, which is best performed on dot plots or histograms, can be thought of as placing a line somewhere in the middle of the data points, and determining whether one side mirrors the other. For example, if you could place a line in the middle of your data points on a piece of paper, and you folded it along the line you have drawn, would they roughly fit over each other?

Modality. The modality of data refers to the most frequent points in your data set. For the graph of male and female salaries (Figure 4.12), the male graph looks roughly unimodal – in other words, there is one point on the graph of male salaries that is more frequent than the rest (this sits around the $100,000 mark). If there were two clear modal values, and the data spread looked more like a camel's back with two humps, we would say the data has a bimodal distribution. Three humps would be trimodal.

Outliers. An outlier is a data point that appears to be far away from the rest of your data points. There is sometimes the temptation to remove outliers from a data set, which should be avoided! It is much better to ask questions about that data point. Why is it so far away from the others? Could there have been a mistake in reading the data? Or is the outlier of interest? Removing a perceived outlier can have big implications for any statistical analysis you perform on the data set and might not give you an accurate representation.

· · · · · · · · Develop your skills! 4.5 ·

Using Figure 4.12, answer the following questions:

1   Do you think the distribution of data for females is trimodal? Explain your answer.
2   Is there anything that surprises you about the side-by-side plots in Figure 4.12?
3   If a larger sample were taken for the female group, do you think it would resemble the male group – would the distribution be similar? Explain your answer.

· ·

Try to avoid:

1   Automatically assuming data points are outliers, and don't assume you can safely remove them or ignore them.
2   Making big claims about small samples, especially if you are trying to extend your findings to larger groups or populations.

4.2.4 Cross-tabulations and one/two-way tables of counts

Using **tables** is a common way of presenting data, especially in the social sciences. When a table presents data for one, and only one, categorical variable, it is called a **one-way table**. A one-way table is the tabular equivalent of a bar chart. Like a bar chart, a one-way table displays categorical data in the form of frequency counts and/or relative frequencies.

The data summary below leads onto a one-way table of counts. It's called one-way since it only deals with one variable. The variable of interest here is Region.

Primary variable of interest: Region (categorical)

Total number of observations: 29,824

You can see in Table 4.4 that the number of counts (which is the number of schools) for each region has been recorded, and then these have been added up to form a total (which is 29,824 for this data set). Then this value is used to help calculate the percentage of counts of each region, in relation to the total count for all regions.

Table 4.4   One-way table of counts

Region	1	2	3	4	5	6	Total
Count	11,570	9220	2947	2489	2270	1328	29,824
Percentage (%)	38.79	30.92	9.88	8.35	7.61	4.45	100

In a **two-way table of counts**, you will see that there are now two variables included (Table 4.5). These kinds of tables are extremely useful and can be created to add things like percentages (which use the counts in the table, along with the row and column totals to help achieve this). The example in Table 4.5 looks at the crime levels in San Francisco, 2014. The crimes used in this table include Theft and Assault, which were recorded before (AM) or after noon (PM).

Table 4.5   Two-way table of counts: crime levels in San Francisco, 2014

	Theft	Assault	Total
AM	23,275	6,521	29,796
PM	56,082	14,122	70,204
Total	79,357	20,643	100,000

····· Develop your skills! 4.6 ·· · ·····················

Using the data in the Excel spreadsheet 'Banks in USA 2020', which you can access as part of the online resources (https://study.sagepub.com/jonesstatsliteracy), answer the following questions:

1   Construct a one-way table of counts for the variable 'Headquarters location', for the first 20 largest banks in the USA in 2020.
2   Which headquarter location had the highest count, from the first 20 largest banks in the USA, in 2020?

···

Key points to remember

1   Be careful how you identify extreme data point values are outliers. Context-specific information is needed to ask if an extreme data point really is an outlier. This can be gained from subject experts, or potentially from the people who collected the data.

(Continued)

2   Try not to remove perceived outliers from plots.
3   Explaining relationships in scatter plots as one variable causes another variable to increase or decrease can be dangerous and misguided. Causal relationships can be ascertained from using inferential techniques, not by describing the relationship observed in a scatter plot.
4   Use the guidance presented in this chapter to fully explain plots, using the correct terminology.

. . . .

References to support this chapter

Bach, B., Kerracher, N., Wm, K., Carpendale, S., Kennedy, J. and Riche, N. (2016) Design patterns for data comics. *Proceedings of ACM Conference on Human Factors in Computing Systems (CHI), 2016*. https://hal.inria.fr/hal-01256099/document

Bach, B., Wang, Z., Farinella, M., Murray-Rust, D. and Riche, N. (2018) Design patterns for data comics. *Proceedings of ACM Conference on Human Factors in Computing Systems (CHI), 2018*. https://dl.acm.org/doi/10.1145/3173574.3173612

FBI (2019) Bank Crime Statistics 2019. https://www.fbi.gov/file-repository/bank-crime-statistics-2019. pdf/view (accessed 23 August 2021).

'List of largest banks in the United States' (2021) *Wikipedia*. https://en.wikipedia.org/wiki/List_of_largest_banks_in_the_United_States (accessed 23 August 2021).

. . . .

5

THE LANGUAGE OF STATISTICS AND STATISTICAL INFERENCE

An overview of this chapter

Building on the previous chapter, emphasis will be placed on key words and common notations that are used in presenting and communicating interesting patterns and trends in data (such as test statistics, *P*-values, and statistical and practical significance), as well as the results from statistical analyses that may have been undertaken. You will be encouraged to see the distinction between using certain words in statistics, reinforcing the shades of grey analogy in the previous chapter. This chapter will help you to develop your statistical literacy skills, using a range of real-world examples. Examples will be drawn from scenarios that involve the interpretation of confidence intervals, an introduction to statistical inference, and an overview of the difference between practical and statistical significance.

Resources to support this chapter

This chapter draws on a range of sources, assisting your learning with the interpretation of confidence intervals, an introduction to statistical inference, and an overview of the difference between practical and statistical significance. In this chapter, we use a range of real-world examples to show you how to analyse different variable types. You can build your skills by trying the suggested activities, helping you to develop your skills in being able to interpret and create a confidence interval, and interpret *P*-values.

Resource	Date accessed	Location
SENSITIVE DATA: Protein vaccines 101 and rising temperatures – the week in infographics, *Nature*, 2021	21 April 2022	https://www.nature.com/articles/d41586-021-03421-6
Scientists' grasp of confidence intervals doesn't inspire confidence, *Science News*, 2014	21 April 2022	https://www.sciencenews.org/blog/context/scientists-grasp-confidence-intervals-doesnt-inspire-confidence

(Continued)

(Continued)

Resource	Date accessed	Location
SENSITIVE DATA: The Little Albert experiment: A closer look at the famous case of Little Albert, Verywell Mind, 2019	21 April 2022	https://www.verywellmind.com/the-little-albert-experiment-2794994
The distribution of *P*-values in medical research articles suggested selective reporting associated with statistical significance, *Journal of Clinical Epidemiology*, 2017	21 April 2022	https://www.sciencedirect.com/science/article/abs/pii/S0895435616308381
Does everything has to be decided by a p-value? Why scientists have been very vocal about the misuse of p-value? *The Startup*, 2020	21 April 2022	https://medium.com/swlh/demystifying-the-p-value-f60f2ccfab27

The websites referred to in all the activities are provided as part of this book's online resources. You can find them at **https://study.sagepub.com/jonesstatsliteracy**.

5.1 Statistical inference and defining parameters

We are often interested in making statements about a population, from a certain country, or perhaps part of a country, but usually do not have access to all the population's data (Chapter 3). We want to find out something about a population such as a population average, for example a population's average time spent online per day (the true value of a parameter).

Often when data are being collected we do not have access to complete population-level data, so we need to use samples to try and say something meaningful about the population they came from (think back to Chapter 3 when we covered sampling and population-level data). Since we are using samples, this suggests there will be some level of uncertainty about what we can say about the population the sample has been taken from. So, our goal is to obtain information about some clearly defined *population*, for example, the height of female giraffes in London Zoo.

A common term used in statistics is **parameter**, which is a numerical characteristic of a population or distribution, such as a population mean. Defining parameters in statistics is of utmost importance as it provides a clear and precise framework for analysing and interpreting data. Parameters serve as the numerical values that characterise a population, and they play a crucial role in various statistical analyses and decision-making processes. By precisely defining parameters, we can establish the foundation for hypothesis testing, estimation, and making inferences about the population based on sample data. Moreover, well-defined parameters facilitate the comparison of different data sets and enable researchers to draw meaningful conclusions from their findings. Without properly defined parameters, statistical analyses can become ambiguous and unreliable, hindering the accurate understanding of data, and impeding the advancement of knowledge in various fields. Thus, the importance of defining parameters in statistics cannot be overstated, as it ensures rigour, clarity and soundness in statistical analyses and enhances the credibility of research outcomes.

An **estimate** is a *known* quantity calculated from (sample) data to estimate an *unknown* parameter; for example, a sample mean can be used to estimate the unknown population mean. The use of estimates in statistics is of paramount importance as it allows researchers to draw meaningful conclusions and make informed decisions based on limited or incomplete data. In many real-world scenarios, it is impractical or impossible to collect data from an entire population, making it necessary to rely on samples. By using estimates, we can make reliable inferences about population parameters based on the information obtained from the sample. Estimates provide an approximation of the true values of population parameters, such as means, proportions, or variances, and they are accompanied by measures of uncertainty, such as confidence intervals or margins of error. These estimates enable researchers to quantify the level of confidence in their findings and assess the reliability of their conclusions. Moreover, estimates allow for generalisation, as they provide insights into the characteristics of the population beyond the observed sample. By using estimates, we can effectively analyse data, make predictions, and guide decision-making processes in a wide range of fields, including business, medicine, social sciences, and public policy.

The process of using sample data to make useful statements about a population is called **statistical inference**, for example when using sample data to estimate an unknown population mean. But an estimate of an unknown mean will often not be enough by itself, and you will need an idea of the spread around that mean. You could also be looking at a sample median, and you could also be dealing with differences between sample means, a single proportion from a sample, or differences between proportions from two different samples.

The following points present several important ideas to think about in relation to an interval estimate:

1   Randomly sample from the population and calculate an estimate of the true unknown parameter value. But because all estimates are *just estimates,* there's still *uncertainty* about the true value of the parameter. We say *there is uncertainty in the estimate.*

2   To account for this uncertainty, we give a range or interval of plausible values (an *interval estimate* or *confidence interval*) rather than a single value (*a point estimate*) for the unknown true parameter value.

3   How do we build this interval? There are two methods that can be used: bootstrapping or the normality-based method.

SENSITIVE DATA EXAMPLE: Climate anxiety

Global warming is often a hot topic in the media, the news, and yearly conferences to address related issues. The following questions are based on this topic.

Figure 5.1 Increasing levels of desertification are likely to be due to climate change and rising surface temperatures

Source: Photo by Peter Burdon on Unsplash

Climate over 24,000 years

As world leaders and scientists discuss how to avert catastrophic climate change at the COP26 meeting in Glasgow, this striking graph puts our current situation in perspective. It shows changes in the Earth's global temperature over the past 24,000 years.

A team of researchers combined climate models with proxy data – obtained indirectly through palaeoclimate records such as ocean sediments – to work out the evolution of the difference in the global mean surface temperature relative to the average for the pre-industrial period of the past millennium (1000–1850). The shaded area represents a 95% confidence interval.

If confirmed through further studies, write the authors of this News & Views article, the results imply that modern warming is extraordinary compared with that of the past 10,000 years – adding weight to a similar conclusion made in the most recent report from the Intergovernmental Panel on Climate Change.

Carbon dioxide emissions on the up

The abrupt decline in global carbon dioxide emissions during the COVID-19 pandemic, caused by government-mandated lockdowns, will be all but erased by the end of this year, a consortium of scientists reported this week. It predicts that carbon emissions from burning fossil fuels will rise to 36.4 billion tonnes – an increase of 4.9% – in 2021 compared with last year. The rapid rebound, driven in part by the increasing demand for coal in China and India, suggests that emissions will begin to rise anew next year without substantial government efforts to bend the curve.

Look at the article, 'Protein vaccines 101 and rising temperatures – the week in infographics', from *Nature*, published in 2021 (https://www.nature.com/articles/d41586-021-03421-6) or read the excerpt on p. 54. Answer the following questions related to this article:

1   Read the first part of the article: 'Climate over 24,000 years'. What is the parameter of interest the team of researchers are investigating in relation to the combined climate models they are using?
2   Read the second part of the article: 'Carbon dioxide emissions on the up'. What is the parameter of interest in this section?

Figure 5.2   Milford Sound, New Zealand, ecological parameters
Source: Photo by Aneta Harmannová on Unsplash

Parameters are everywhere, all around us, in the natural, social and physical world. They form an integral part to many investigations, which may require a clear explanation with regard to the parameter being researched. Look over the following scenarios and see if you can identify the parameter being investigated:

3   An ecological research team decide to look at factors that can influence the vertical growth of plants in Milford Sound, New Zealand.
4   A psychologist begins to investigate personality traits that are associated with likeability.
5   A physicist is asked to give a talk on the mean time it takes for stars to explode in the universe.
6   A social scientist applies for research funding to ascertain factors caused by Covid-19 that can potentially lead to financial deprivation, in the UK.
7   A market analyst is asked to conduct research into online shopping behaviours over the last 5 years, concentrating on mean spend per year.

. . . .

Try to avoid:

1   Confusing the two different methods for generating a confidence interval. This will be explained in more detail in the following section.
2   Assuming an estimate is the true value of a parameter of interest from a population. Remember estimates are just that, estimates, and are usually calculated from a sample that has been taken from a population of interest.

5.2 Confidence intervals: Bootstrapping and normality-based

As mentioned in earlier parts of this chapter, we are often interested in making statements about a population, from a certain country, or perhaps part of a country, but usually do not have access to all the population's data (Chapter 3). We want to find out something about a population such as a population average, for example, a population's average time spent online per day (the true value of a parameter).

So, we need to take obtain a random sample from the population and calculate the sample average (an estimate). All estimates are just estimates – there's still uncertainty about the population's average time spent online per day. Looking at the world using data is like looking down through the sea and trying to make sense of the seabed, where ripples in the waves can distort our image: what we see is not quite the way it really is.

Figure 5.3 Ripples in the sea

Source: Photo by Marissa Rodriguez on Unsplash

To try to account for this sample average possibly being 'not quite right', we use an interval (or range) around this sample average. A **bootstrap confidence interval** is one method of enabling us to do just that.

5.2.1 Bootstrap confidence interval: How do we build one?

Software programs such as the Visual Inference Tool (VIT) Online (https://www.stat.auckland.ac.nz/~wild/VITonline/) will do this process for you. iNZight lite will also provide you with confidence interval data and is also loaded with data set examples. Such programs calculate a bootstrap confidence interval as follows (see also Figure 5.4):

1   From the original sample, randomly sample (with replacement) until the size of
    the re-sample equals the original sample size. So, for example, if there are 10 data
    points in the sample, this process will randomly take one of those data points,
    put it back into the sample, and then take another data point, put it back into
    the sample, and do this a total of 10 times (i.e. to equal the original sample size).
2   These values are then used to calculate an estimate (like a mean or median
    height, or weight).
3   This is done over and over (say, 1000 times) to obtain 1000 estimates.
4   The central 95% of these 1000 estimates are used to find the upper and lower limits
    of an uncertainty interval or interval estimate (bootstrap confidence interval).

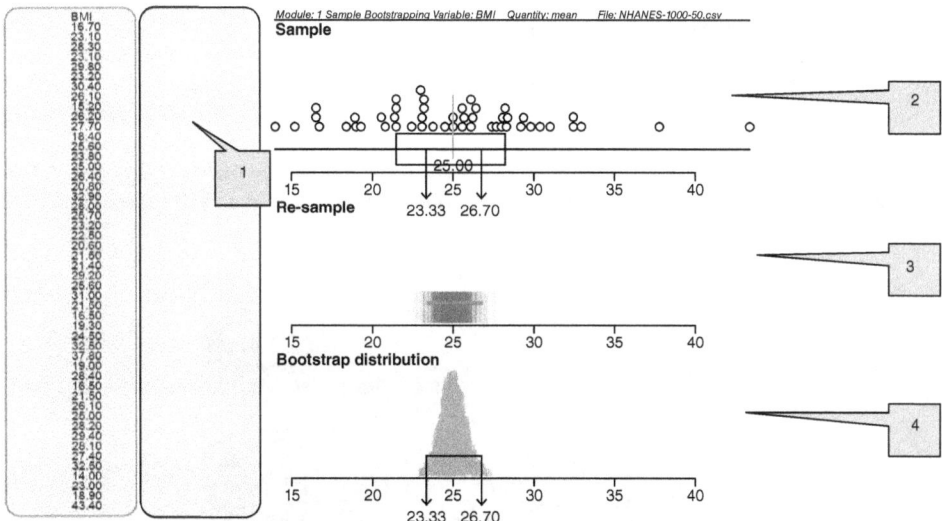

Figure 5.4 Bootstrap confidence interval construction using VIT online

The variable being displayed in Figure 5.4 is based on body mass index (BMI) values
from a sample of 49 individuals, taken from a larger population. The numbers on the
left of the figure (marked 1) are the original 49 sample data points or values that are used
to generate the bootstrap confidence interval. The top right-hand panel in the figure
(marked 2) represents a single bootstrap sample: samples are taken from the values on
the left of the figure, with replacement, until this bootstrap sample is the same size as
the original sample (49). We record the estimate of the mean, and repeat this process
1000 times. The middle right-hand panel (marked 3) shows all the 1000 estimated mean
values that have been generated in this way. The bottom right-hand panel (marked 4)
shows the final step, which generates the bootstrap confidence interval, which is a range
of plausible values for the true population parameter, which in this case will be the BMI
value for the population the sample has come from. The result is a bootstrap mean of
25, with a 95% confidence interval from 23.33 to 26.70. The pyramid type shape in this
panel is called the bootstrap distribution.

Note:

1   The distribution of estimates from the re-samples (bootstrap samples) is called a bootstrap distribution (in Figure 5.4 this is marked 4).
2   Our confidence in a particular bootstrap confidence interval comes from the fact that this method produces an interval which contains the true value of the parameter a very high percentage of the time (95%).
3   Interpretation of the confidence interval: it's a fairly safe bet that the true value of the parameter is one of the values in this confidence interval.
4   A bootstrap confidence interval for a parameter can be thought of as a range of plausible values for the true parameter value.

· · · · · · · **Develop your skills! 5.2** ·

SENSITIVE DATA EXAMPLE: Climate anxiety

We know that global warming is often a hot topic in the media, the news, and yearly conferences to address related issues. The following questions are based on this topic.

Figure 5.5   Pollutants being released into the atmosphere

Source: Photo by Chris Leboutillier on Unsplash

Climate over 24,000 years

As world leaders and scientists discuss how to avert catastrophic climate change at the COP26 meeting in Glasgow, this striking graph puts our current situation in perspective. It shows changes in the Earth's global temperature over the past 24,000 years.

A team of researchers combined climate models with proxy data – obtained indirectly through palaeoclimate records such as ocean sediments – to work out the evolution of the difference in the global mean surface temperature relative to the average for the pre-industrial period of the past millennium (1000–1850). The shaded area represents a 95% confidence interval.

If confirmed through further studies, write the authors of this News & Views article, the results imply that modern warming is extraordinary compared with that of the past 10,000 years – adding weight to a similar conclusion made in the most recent report from the Intergovernmental Panel on Climate Change.

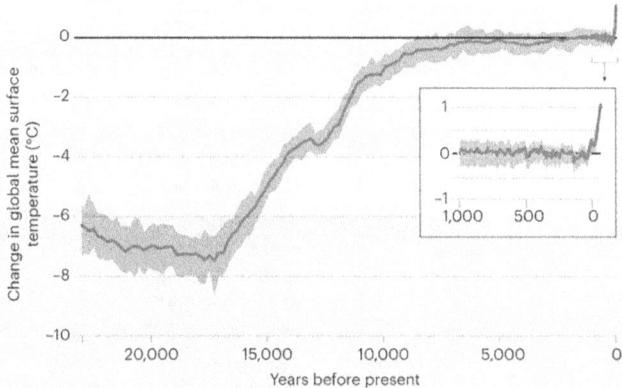

Figure 5.6 Rising temperatures graph
Source: Protein vaccines 101 and rising temperatures – the week in infographics, 2021, Springer Nature, https://doi.org/10.1038/d41586-021-03421-6

Look at the article, 'Protein vaccines 101 and rising temperatures – the week in infographics', from *Nature*, published in 2021 (https://www.nature.com/articles/d41586-021-03421-6) or read the excerpt on pp. 58–9. Answer the following questions related to this article:

1   Read the first part of the article, 'Climate over 24,000 years', and look at the figure concerning the change in global mean surface temperatures, over last 25,000 years. What does the red shaded part around the solid red line represent?
2   Still focusing on the figure you looked at for Question 1, why do you think the past 1000 years have needed to be enlarged?
3   Do you think it would be appropriate to use a bootstrap method to generate a confidence interval for the change in global mean surface temperature over the last 25,000 years if we had access to the relevant data?

. . . .

Try to avoid:

1   Misinterpreting what a confidence interval represents. Remember it is a range of plausible values for an unknown parameter of interest (like weights, heights, proportion of people who like Oasis etc), which is generated by taking a sample from a population.
2   Assuming that confidence intervals are 100% accurate or give you the true population parameter value.

5.2.2 Normality-based confidence interval

Now that we've looked at bootstrap confidence intervals, let's turn to normality based confidence intervals. Often in statistics we are interested in finding parameter values, from a sample average. Table 5.1 shows common notations used to represent what we are interested in, from a sample. It also includes examples applicable to the notations used in the table.

Table 5.1 Parameter and sample notation used in statistics

Description	Parameter	Example	Estimate	Example
Single mean	μ (called mu)	*Population mean weight of children in Buenos Aires*	$\bar{x}$(called x bar)	*Sample mean weight of children in Buenos Aires*
Single proportion	p	*Population proportion of people who like to gamble, who live in Las Vegas*	$\hat{p}$	*Sample proportion of people who like to gamble, who live in Las Vegas*
Difference in 2 means (from 2 independent samples, which means the samples are selected randomly and the observations (i.e., data points) from one group do not depend or have any bearing on observations from the other group)	$\mu1 - \mu2$ (mu1 – mu2)	*Population Height of females in Wales – Population Height of males in Wales*	$\bar{x}1 - \bar{x}2$ (x bar 1 – x bar 2)	*Sample Height of females in Wales – Samples Height of males in Wales*
Difference in 2 proportions	$p1 - p2$	*Proportion of people who like french fries in Italy – Proportion of people who like pizza in Italy*	$\hat{p}1 - \hat{p}2$	*Sample of people who like french fries in Italy – Sample of people who like pizza in Italy*

Normality-based (or norm-based) confidence intervals involve a similar process to the bootstrapping technique, with the important difference that data (which can be variables measuring heights, weights, test scores from an exam) are assumed to be normally distributed. The normal distribution is characterised by the following properties:

1   The distribution is symmetrical and bell-shaped (see Figure 5.7)
2   It is centred at the mean, denoted by μ
3   The mean, mode and median all have the same value
4   The total area under the curve is 1

Samples are then taken from the population (i.e., different samples) and not replaced. These sample estimates are then used to generate a norm-based confidence interval. This

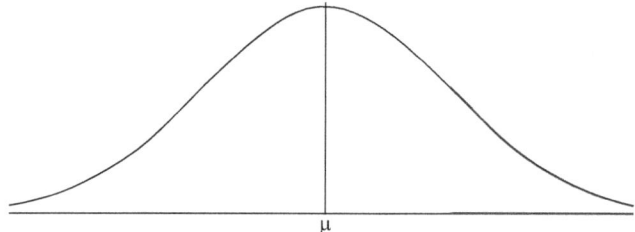

Figure 5.7   The normal distribution

interval will have the same form as a bootstrap confidence interval (i.e., a lower and upper limit), which gives us our range of plausible values for the true population parameter.

Norm-based confidence intervals can be reported either as a lower and an upper limit, or by giving the estimate plus or minus the width of the confidence interval divided by 2. This is because we are assuming the data are normally distributed and therefore symmetrical, so the estimate will always be in the middle of our confidence interval. Often political polls like to present estimates (e.g., voting percentages in a presidential campaign) using plus or minus half the confidence interval width, which is often referred to as the margin of error.

Since bootstrap confidence intervals can use median values as the estimate, this does not assume the distribution of data is normal, and therefore does not assume the distribution of data is symmetrical. So bootstrap confidence intervals are often reported using the lower and upper limits.

Both are valid techniques and can be used to give us confidence that our intervals are picking up the true population parameter most of the time (i.e., 95%), using samples (for norm-based approaches), or one sample (for bootstrap approaches) from the population of interest.

········ Develop your skills! 5.3 ··· ···

The following questions will help you to practise your skills in producing a confidence interval, along with their interpretation.

Figure 5.8 Family
Source: Photo by Tyler Nix on Unsplash

We want to find out about the median weight of the Jones family (yes, my family – let's pretend this is something of interest to us for the moment) and we can gain access to 47 weights. Now let's say we would rather take a random sample of 10 weights from the Jones family, which will be quicker and therefore cheaper (if we are hiring a research assistant to collect the data). The 95% confidence interval (using a norm-based method) for this random sample of 10 weights is 83.6 ± 8.11

(Continued)

1   What is the width of the confidence interval?
2   What are the lower and upper limits of the confidence interval?
3   Can we be certain that this confidence interval contains the true population parameter, from which the sample has come from? Explain your answer.
4   If we calculated a bootstrap confidence interval and our estimate involved using the median weight from the sample of 10, and then we were presented with the lower and upper limits, would we be able to find out what the median value is? Explain your answer.

Read the article, 'Scientists' grasp of confidence intervals doesn't inspire confidence' (at https://www.sciencenews.org/blog/context/scientists-grasp-confidence-intervals-doesnt-inspire-confidence) and answer the following questions:

5   Recall the following statement in the article:

> Ordinarily you might see a confidence interval expressed something like this:
>
> The average weight loss for people on the new miracle drug was 4.6 pounds, with a 95 percent confidence interval of 2.2 pounds to 6.9 pounds.

Assuming a norm-based method has been taken, what is the margin of error?

6   Concerning the 95% confidence interval presented in Question 5, the article goes on to say:

> But what does that really mean? Some people would say it means if you did the experiment 100 times, the average (mean) you got would be within that 2.2–6.9 range in about 95 of the trials.

Why is this interpretation incorrect?

7   Do you think the title of this article, 'Scientists' grasp of confidence intervals doesn't inspire confidence', is fair? What evidence does the author present to validate the claim they make in their title?

. . . .

Try to avoid:

1   Mixing up the notation used for population values (like population means) and sample values (such as sample means). Notation in statistics is very important and you need to make sure you have these clear in your head, because they outline and define what you are looking for or referring to. Making a clear distinction between the two also enables you to be clear about what data you are working with, and where the data have come from. It also enables you to make sensible inferential statements.
2   Forgetting that the variable or parameter of interest you are generating a norm-based confidence interval for is normally distributed. This is an important consideration for using norm-based methods to calculate a confidence interval, and it is an assumption that needs to be judged by looking at a dot plot of the data (i.e., assumptions that the data is normally distributed – bell-shaped

distribution and roughly symmetrical, the distribution is centred at μ (i.e., the mean) and the mean, mode and median all have the same value).

5.3 More on statistical inference

Often when we want to make an inferential statement, using a sample that has come from a larger population, the sample needs to be randomly selected, to help mitigate against any biases (recall Chapter 3, where we looked at sampling and non-sampling errors). When we take sample data like this and want to say something about the population the data have come from (like we did in Develop Your Skills 5.3, in Questions 1–4), we call this kind of inference **sample-to-population inference** (Table 5.3). If, however, we are looking at experimental data, which involved random allocation of a treatment of some kind to groups in an experimental set-up, we may want to make a causal inference (e.g., when we change one variable, say, temperature, it *causes* a change in another variable, which could be rate of sweating in humans), which we call **experiment-to-causation inference** (Table 5.3). Chapter 8 explores experiments and observation studies in more detail. Chapter 9 will also explore these concepts of causal inference in more depth, when we look at how to read quantitative journal articles.

Table 5.3   Inference types used in statistics

Random processes involved with data production	Inference (justified by study design)
Random allocation of units to treatment groups (an experiment)	Experiment-to-causation inference
Random sampling of units from a population or populations	Sample-to-population inference

Try to avoid:

1   Confusing the two types of inferential statements presented in this section. The type of inference you are making will determine the way you present your statements to describe sample data, and how you use this to potentially generalise to the population the sample came from.

5.4 Hypothesis testing

Hypothesis testing is often used in statistics to try and make a statement about a variable or variables of interest. The main components of **hypothesis testing** are:

1   The hypotheses
2   The test statistic

3   The *P*-value

4   *P*-value interpretation

Hypothesis testing is used in inferential statistics, and can include:

1   Experiment-to-causation inference

2   Sample-to-population inference

Let's say a group of researchers conduct a study to check out a hunch or an idea that they may have about a certain phenomenon, in the natural, social or physical world. The idea gives rise to a research **hypothesis** which the researchers try to establish as being true.

Hypothesis tests often involve two competing statements called the null hypothesis and the alternative hypothesis. The **null hypothesis**, denoted by H_0, is the statement that we test. We determine how much evidence we have against H_0.

The null hypothesis usually takes a neutral or doubtful point of view: the research-er's hunch is unconvincing, there is nothing new or interesting happening, there is no effect. In other words, the results from, say, a well-designed experiment are just due to chance, and have nothing to do with any factors, that is to say, any changes we see in one variable are not being caused by the researchers changing another variable.

In most situations the researcher hopes to disprove or reject H_0. We can never show or prove that H_0 is true, because when you are able to reject the null hypothesis, you cannot rule out chance completely: there is always an element of chance that can influence the results you see in an experiment like the one described in Table 5.4 (see below). However, you can (if you get a *P*-value that is smaller than 0.05 (i.e., 5%)), rule out chance acting alone and fend off suggestions that any differences you see in data from an experiment are purely due to luck of the draw results.

The **alternative hypothesis**, denoted by H_1 and sometimes called the research hypothesis, is usually written to explain that something is happening, there is a differ-ence or an effect, there is a relationship. In most situations the researcher hopes to give support to H_1 by showing that H_0 is not plausible.

For example, suppose we are interested in looking at the effectiveness of two diets in helping to lose weight in a random sample of children, and they are randomly assigned to two groups, with one being given diet A, and the other group given diet B. Our hypotheses would be as follows:

H_0: Diet A and diet B have the same effect, or there is no difference in weight

H_1: Diet A and diet B do not have the same effect, or there is a difference in weights in the two groups

In this next example let's say we are interested in the amount of TV children watch, and whether it can harm their mental health. Our hypotheses would be as follows:

H_0: TV does not affect children's mental health

H_1: Too much TV can be harmful to children's mental health

In a final example, let's say a research group is interested in global warming and sea levels. Their hypotheses might be as follows:

H_0: Global warming and sea levels are not associated, or they are not related, or they are independent

H_1: Global warming and sea levels are related, or are associated, or they are dependent

So, in summary, the researcher has an idea (which can be based on existing knowledge or data from other research) which takes the role of the alternative hypothesis (i.e., the research hypothesis), which then must have a null hypothesis to test against the alternative hypothesis. They then test the 'plausibility' of the null hypothesis by asking 'does the evidence (data) suggest that the null is simply not plausible?' If so, then their alternative hypothesis gains support.

Now let's go through one example (Table 5.4), that takes us through the stages of hypothesis testing. In this example let's assume it is a well-designed experiment, where one group have been randomly allocated one bar of chocolate to consumer in 1 minute, and their heart rates are measured before and after the consumption of chocolate.

Table 5.4 Deductive method involving hypothesis testing approach

Example study	Does the consumption of chocolate influence heart rate in humans?
Null hypothesis: H_0	Chocolate consumption does not affect heart rate (in other words, any changes in heart rate we see are just luck of the draw results, i.e., chance is acting alone)
Alternative hypothesis: H_1	Chocolate consumption does affect heart rate (in other words, any changes in heart rate we see could be caused by the consumption of chocolate)
Data	The difference between mean heart rates of the group who ate the chocolate and the group who did not eat the chocolate is 22.6 beats per minute
P-value If the hypothesis is true, then what are the chances of the data scenario occurring?	An observed difference of 22.6 beats per minute is highly unlikely when chance is acting alone. (This is when you would look at the P-value.) If the P-value is smaller than 0.05 (i.e., 5%) then we can reject H_0
Interpretation/Conclusion Is the hypothesis plausible?	Therefore, *chance was probably not acting alone. The observed difference (of 22.6 beats per minute) was (at least partially) due to the consumption of chocolate.*

Another group are given no chocolate, and their heart rates are measured before and after a minute has passed.

The table includes reference *to P*-values, which we'll look at more in Section 5.5.

Develop your skills! 5.4

SENSITIVE DATA EXAMPLE: Child abuse

The series of questions below are based on a famous psychology study titled, the Little Albert experiment. The questions will help you to further develop your skills in being able to write hypotheses.

Figure 5.9   Picture of a baby

Source: Photo by Christian Bowen on Unsplash

Look at the following webpage: 'The Little Albert experiment: A closer look at the famous case of Little Albert' (at https://www.verywellmind.com/the-little-albert-experiment-2794994) and answer the following questions:

1   The Little Albert experiment is a famous psychology study that investigated condition processes in humans. After reading the first paragraph at the top of the webpage, see if you can identify what the parameter of interest is here.

2   Write a null and alternative hypothesis for this experiment.
3   Are there any ethical issues you can think of related to this experiment?

. . . .

Try to avoid:

1   Mixing up the null and alternative hypotheses. It is really important that you get
     these the right way around, and that you can construct your own hypotheses
     from a scenario that could be an experiment or some sort of phenomenon in the
     natural, social or physical world.
2   Forgetting that in hypothesis testing, it's the null hypotheses we test and look for
     evidence against (determined by looking at a P-value – more on this in the next
     section).

5.5 Test statistics, significance levels and P-values

5.5.1 What is a test statistic?

You may have come across test statistics at some point in your studies already and may
have some idea that they are linked to generating P-values. This section will go into
more detail into the use of test statistics, and why they are useful in being able to interrogate data.

Test statistics measure the difference between what we see in the data and what
we would expect to see if the null hypothesis, H_0, were true. A test statistic is a summary
statistic that we use to evaluate the hypotheses in a test. That is, the test statistic is what
we use from the data as evidence against the null hypothesis.

The types of variables we are measuring (e.g., whether they are categorical or ratio)
and the different numbers of groups we are looking at will influence the type of inferential test we conduct, and the type of test statistic we work with. The test statistic is a
value that helps us to get a P-value, which tells us how much evidence we have against
the null hypothesis, H_0. Some common inferential statistical tests that lead to the creation of a corresponding test statistic include:

1   The Student's t-test (which involves the t-statistic or t_0)
2   A one-way analysis of variance, or one-way ANOVA (which is also known as the
     F-test)
3   The chi-square (or χ^2) test

Many software packages (such as SPSS or R) can calculate test statistics and P-values, if
you have the data in a form that they can read. It is great we have software that can do
this; however, the interpretation of P-values is something you need to learn and be able
to do, because most software will not tell you this all of the time. Usually, the *larger* the

test statistic, the *smaller* the P-value, which means that there is a *greater amount of evidence against* the null hypothesis, H_0.

5.5.2 What is the P-value?

The **P-value** is the conditional probability of observing a test statistic as extreme as or more extreme than that observed, given that the null hypothesis, H_0, is true. This definition is not the easiest to get your head around! So let's focus on what it does, which is very important.

The P-value gives us an idea of how likely (or unlikely) it is to get the results we did just by chance (when the null hypothesis is true). It measures the strength of evidence against the null hypothesis, H_0. The smaller the P-value, the stronger the evidence against H_0.

When we interpret the P-value, which we can obtain from a test statistic, Table 5.5 can be used as a guide to tell us the strength of evidence we have against the null hypothesis, H_0.

Table 5.5   P-value interpretations for evidence against the null hypothesis

P-value	Evidence against the null hypothesis, H_0
> 0.10 (greater than 10%)	None
≈ 0.07 (approximately 7%)	Weak
≈ 0.05 (approximately 5%)	Some
≈ 0.01 (approximately 1%)	Strong
≤ 0.001 (below and including 0.1%)	Very strong

Fixed-level statistical significance (i.e., when researchers or news articles look at P-values) involves a cut-off point, which is usually 0.05 (5%). Table 5.6 provides some guidance as to how this is achieved.

Table 5.6   Significance level testing cut-off points

P-value	Test result	Response
< 0.05	Significant	Reject H_0 in favour of H_1
> 0.05	Non-significant	Do not reject H_0

Testing can be done at other levels of significance (e.g. at the 1% level, with a P-value less than 0.01), which is common in certain fields such as pharmaceutical research, especially for drugs that can affect the human heart.

The following series of questions will hone your skills in being able to spot and interpret *P*-values, which are presented in a journal article.

Figure 5.10 Scientist carrying out medical research
Source: Photo by National Cancer Institute on Unsplash

Below is an excerpt from a 2017 article, 'The distribution of *P*-values in medical research articles suggested selective reporting associated with statistical significance', from the *Journal of Clinical Epidemiology* (at https://www.sciencedirect.com/science/article/abs/pii/S0895435616308381). Don't worry if you can't understand the whole article – answers to the questions that follow should be easy enough for you to find, based on the content presented in this section of the chapter, which is reprinted with permission from Elsevier.

Abstract

Objectives

Published *P*-values provide a window into the global enterprise of medical research. The aim of this study was to use the distribution of published *P*-values to estimate the relative frequencies of null and alternative hypotheses and to seek irregularities suggestive of publication bias.

(Continued)

Study Design and Setting

This cross-sectional study included *P*-values published in 120 medical research articles in 2016 (30 each from the *BMJ*, *JAMA*, *Lancet*, and *New England Journal of Medicine*). The observed distribution of *P*-values was compared with expected distributions under the null hypothesis (i.e., uniform between 0 and 1) and the alternative hypothesis (strictly decreasing from 0 to 1). *P*-values were categorized according to conventional levels of statistical significance and in one-percent intervals.

Results

Among 4,158 recorded *P*-values, 26.1% were highly significant ($P < 0.001$), 9.1% were moderately significant ($P \geq 0.001$ to < 0.01), 11.7% were weakly significant ($P \geq 0.01$ to < 0.05), and 53.2% were nonsignificant ($P \geq 0.05$). We noted three irregularities: (1) high proportion of *P*-values <0.001, especially in observational studies, (2) excess of *P*-values equal to 1, and (3) about twice as many *P*-values less than 0.05 compared with those more than 0.05. The latter finding was seen in both randomized trials and observational studies, and in most types of analyses, excepting heterogeneity tests and interaction tests. Under plausible assumptions, we estimate that about half of the tested hypotheses were null and the other half were alternative.

Conclusion

This analysis suggests that statistical tests published in medical journals are not a random sample of null and alternative hypotheses, but that selective reporting is prevalent. Significant results are about twice as likely to be reported as nonsignificant results.

1   How are the *P*-values presented in the Results section of the article?
2   How do they compare to the descriptions reported in Table 5.5 in this chapter?
3   Why do you think this article is important in relation to the type of research articles it is reviewing?
4   In the Conclusion section of the article, why do you think significant results are more likely to be reported than non-significant results?

. . . .

Try to avoid:

1   Misinterpreting what a *P*-value is, and what it can tell you about results from a study.
2   Assuming that if you do not reject the null hypothesis, then there is no causal relationship between variables of interest in a study.

3   Assuming that if you reject the null hypothesis, then the results you see are purely down to one variable causing a change in another variable. Remember that chance can still be having some effect on results from a study, even if you get statistically significant results (i.e. some/strong/very strong evidence against the null hypothesis).

4   Concluding that statistically significant results (i.e. some/strong/very strong evidence against the null hypothesis) give you the authority to generalise your results to other populations, or all participants in an experiment or study.

5.6 Statistical versus practical significance

As we saw above, a small P-value provides evidence about the existence of an effect or difference. However, it says nothing about the size of that effect or difference. The following two statements help to describe the main difference between statistical and practical significance:

Statistical significance relates to the P-value (i.e., how much evidence we have against the null hypothesis)

Practical significance (practical importance) relates to the size of an effect

The size of an effect is estimated with a confidence interval (which can be generated by using either a bootstrap or norm-based method; see Section 5.2). Whether or not an effect/difference is of practical importance is a consequence of how big it is. Look at a confidence interval when determining the practical significance of an effect.

The size of a statistically significant effect can be so small as to have no practical importance at all, that is, statistical significance does not imply practical significance. To help explain this with an example, imagine the following scenario. A research group sets up an experiment where they randomly allocate participants to two groups; one receives diet A and the other diet B. The diets are aimed at reducing blood glucose levels, which can have important implications for a person's health. The participants' blood glucose levels are measured before and after they are given the different diets. After the experiment, the research team conducts a statistical test and looks at the mean difference in blood glucose levels (i.e. before and after the diets were administered) between the groups, and find that they have statistically significant results (i.e., a P-value less than 0.05). However, when they look at the confidence interval for the difference, the lower and upper limits are close together, and the mean difference is small, which results in the research team concluding that the difference in blood glucose levels is of no clinical importance. So even if you can get a statistically significant result, it might be of little or no practical importance.

The following series of questions will help you to further develop your skills in being able to spot and interpret *P*-values, which are presented in a journal article.

Figure 5.11 Making a hypothesis
Source: Photo by Mark Fletcher-Brown on Unsplash

The following is an excerpt from the article, 'Does everything has to be decided by a p-value? Why scientists have been very vocal about the misuse of p-value?' (at https://medium.com/swlh/demystifying-the-p-value-f60f2ccfab27). Answer the following questions, related to this article.

Introduction

The most widely practised methodology for determining statistical significance during null hypothesis significance testing (NHST) is by using p-value. For the uninformed, NHST helps scientists to extrapolate the findings from your sample data to population (you would always need to find if your sample results are indeed generalizable!). The last step in a typical NHST is the determination of statistical significance. The idea behind this is to determine if your sample results are statistically significant or not. This helps in extrapolating the results to population level. Statistical significance is quantitatively determined by p-value. This is why it is a holy grail in research. Reporting of p-value is very common in scientific literature. In an analysis done using PubMed papers published between 1990 and 2015 the authors reported that the use of p-values has increased over the last 25 years and about 78.4% of papers in clinical journals reported p-values which was higher than most other disciplines. It is to be noted that the reporting of p-values in research publications is not a bad practice by any means. It is the misuse of p-value for arriving in categorical or dichotomous conclusions which is what must be curtailed. Using p-value *alone* to conclude whether there is an effect or not is a bad practice for interpreting the results and yet this practice is widely prevalent. One analysis using 791 published research papers published in 5 different journals found that about half of them interpreted the p-values incorrectly by assuming non-significance means no effect absence of evidence is not evidence of absence. Lack of understanding of what exactly is a p-value is rampant among medical fraternity.

1   After you have read through the first five or six sentences, is there anything that stands out in the text? Do you think that there is potentially incorrect any information presented?
2   What are the two main issues the author states towards the end of the article, in relation to P-values?
3   What other type of inferential statistical method do you think the author is referring to (that would help to mitigate against the issues highlighted), when he states that using the P-value alone to conclude whether there is an effect or not is bad practice.

··

·····Develop your skills! 5.7··· ··

SENSITIVE DATA EXAMPLE: Alcohol

Look at the following scenario and decide whether the results are statistically and/or practically significant.

Figure 5.12   Alcohol
Source: Photo by Julia Nastogadka on Unsplash

A research group decides to investigate whether alcohol consumption is related to concentration in humans. Participants are randomly allocated to two groups, where one group are given different amounts of alcohol to drink in 30 minutes, and the other group are given water to drink in 30 minutes. After an hour, participants in each group undertake a concentration test, which measures the time it takes (in seconds) for

(Continued)

participants to spot hazards on a computer screen. The longer it takes the participant to spot the hazards, the lower their concentration levels. After the results are collected, the research team conduct a *t*-test on the mean time difference (in seconds) it took to complete the concentration test, between the group that were given alcohol and the group that were given water. A series of confidence intervals were also generated, using a norm-based method. From previous research and existing knowledge, the research group make the call that a mean time difference above 20 seconds would be considered practically significant, between groups. The results are presented below:

Experiment 1:

Alcohol group: 5 units

Water group: 0 units

P-value = 0.02

The confidence interval for the mean time difference (alcohol group – water group) is (4.5, 12.9). So this interval means that, on average, it took the alcohol group between 4.5 and 12.9 seconds longer than the water group to complete the concentration test.

Experiment 2:

Alcohol group: 10 units

Water group: 0 units

P-value = 0.000016

The confidence interval for the mean time difference (alcohol group – water group) is (13.1, 19.5). So this interval means that, on average, it took the alcohol group between 13.1 and 19.5 seconds longer than the water group to complete the concentration test,.

Experiment 3:

Alcohol group: 20 units

Water group: 0 units

P-value = 0.000035

The confidence interval for the mean time difference (alcohol group – water group) is (21.1, 35.0). So this interval means that, on average, it took the alcohol group between 21.1 and 35.0 seconds longer than the water group to complete the concentration test.

Answer the following questions, based on the results:

1   Which of the three experiments produced statistically significant results (i.e., *P*-values below 0.05?)
2   Which of the three experiments produced practically significant results? Explain your answer.

3   How would you describe the relationship between alcohol units and concentration of the
     participants, from the results in this experiment?
4   Would you be able to calculate the mean time difference between groups, using the confidence
     intervals presented, in each experiment? Explain your answer.
5   Are there any ethical concerns with this study? Try and expand on your answer.

. . . .

Try to avoid:

1   Confusing practical and statistical significance. It is very good practice to state
     both types of significance, when reporting results from a statistical analysis
     performed on data.

5.7 Modelling data (advanced)

When using inferential statistics, the type of model we use to generate a test statistic,
and then the *P*-value, will depend to a large extent on the type of variables we are working with, and the number of groups that we are looking at or comparing.

The use of models in statistics holds great importance as they provide a structured
framework for understanding complex data patterns, making predictions, and gaining
insights into underlying relationships. Models serve as simplified representations of
reality, capturing the key features and dynamics of a system or phenomenon under
investigation. They help us to organise and interpret data by identifying relevant variables, defining their relationships, and quantifying the effects of different factors.
Models can be simple or complex, depending on the nature of the problem and the
available data. They allow for hypothesis testing, parameter estimation, and extrapola-
tion beyond the observed data. Moreover, models provide a means to explore different
scenarios and assess the impact of changes in variables or conditions. They enable
researchers to make informed decisions, evaluate the effectiveness of interventions and
design optimal strategies. Overall, the use of models in statistics enhances our ability to
understand and analyse complex phenomena, and it plays a crucial role in various scientific, economic and social domains, contributing to advancements in knowledge and
evidence-based decision-making.

Table 5.7 presents some of the more common types of inferential statistics used to
model data (which is a way of applying statistical analysis to a data set of interest) from
a study or a sample taken from a population of interest. It also presents the different
kinds of variables of interest you could be analysing/comparing, which links up with
the most appropriate form of inferential statistical test to use. Note that some parts of
this table may not make a lot of sense now; however, the intention is to give you a useful summary for you to come back to later when you have covered these topics in more
detail in your studies.

Table 5.7 Modelling data and real-world examples

Variable/s of interest	Example	Useful displays/ plots/graphs/ tables of the relevant data	Hypotheses	Model	Interpretation and conclusion
			Research (alternative) hypothesis = H_1 Null hypothesis = H_0		Interpret the P-value and decide whether to accept or reject H_0
One numeric variable	Mean salary of department store managers in the UK		$H_0 : \mu = \mu_0$	One sample t-test (mean) *Bootstrap or norm-based CI	
One numeric variable	Mean difference of cholesterol levels in the blood, before and after a low-fat diet	Dot plot and/or box plot Analysing mean/ median	$H_0 : \mu_{Diff} = 0$	Paired data t-test (mean difference) *Bootstrap CI	
One categorical variable	Median proportion of people who believe in climate change in New Zealand	Bar chart Analysing proportions/ percentages	$H_0 : p = p_0$	One sample t-test (proportions) *Bootstrap CI	
One numeric and one categorical	Mean salary of males and females who work in McDonald's in China	Side-by side dot plots and/or box plots on same scale. Comparing or analysing means/ medians; standard deviations	$H_0 : \mu_1 - \mu_2 = 0$ $H_0 : \mu_1 = \mu_2 = \mu_3 = ...$ (H_0 : The means/ medians are the same)	Two independent samples t-test *Bootstrap CI (difference) F-test for one-way ANOVA Randomisation test	

Variable/s of interest	Example	Useful displays/ plots/graphs/ tables of the relevant data	Hypotheses	Model	Interpretation and conclusion
Two categorical variables	Gender and Political Party preference in India	Side by side bar charts of proportions.	$H_0 : p_1 - p_2 = 0$	t-test (check sampling situation)	
		Comparing or analysing proportions/ percentages		(Bootstrap CI difference)	
				χ^2 test of independence	
			H_0 : The distributions are the same	χ^2 test of independence	
			$(H_0$: The proportions are the same)	Randomisation test	
Two numeric variables	Age and income	Scatter plot; trend; association	$H_0 : \beta_1 = 0$	t-test no linear relationship	
Random sample					Sample-population inference
Experiment					Experiment-causation inference
					Results from statistical analysis applies to those in study, can be used to reliably establish a causal link or association between variables of interest
Observational study					Results from statistical analysis applies to those in study, can be used to establish possible causal link between variables of interest

Try to avoid:

1   Confusing variable types
2   Getting overwhelmed by the different types of inferential statistical tests. Take
    one step at a time, and first think about the variables you are using, and the
    levels of measurement within them.

Key points to remember

1   This chapter is probably one of the hardest in the book to get your head around, so try not to
    stress out or worry if it does not sink in the first time you go through the content.
2   The previous chapter to this one is foundational knowledge, so if you are still finding this chapter
    tough, go back over Chapter 4; that should help you tackle this chapter better.
3   Ensure you know the difference between a bootstrap and a norm-based confidence interval,
    and the advantages of using both.
4   Make sure you understand the value of hypothesis testing, as well as the connection between a
    test statistic, a P-value, a confidence interval and what these pieces of information can tell you
    with respect to the relevant null hypothesis.
5   Make sure you are clear in your head about the difference between practical and statistical
    significance.
6   This chapter tackles key concepts in statistics and starts to emphasise the uncertainty that
    statistics attempts to factor in and account for. Modelling data is a process that can help with
    this and enable us to make meaningful statements about sample or experimental data.

References to support this chapter

Cherry, K. (2019) The Little Albert experiment: A closer look at the famous case of Little Albert, *Verywell Mind*. http://www.*verywellmind*.com/the-little-albert-experiment-2794994 (accessed 21 April 2022).

Dennis, R. (2020) Does everything has to be decided by a p-value? Why scientists have been very vocal about the misuse of p-value?, *The Startup*. https://medium.com/swlh/demystifying-the-p-value-f60f2ccfab27 (accessed 21 April 2022).

Nature (2021) Protein vaccines 101 and rising temperatures — the week in infographics. https://www.nature.com/articles/d41586-021-03421-6 (accessed 21 April 2022).

Perneger, T.V. and Combescure, C. (2017) The distribution of P-values in medical research articles suggested selective reporting associated with statistical significance. *Journal of Clinical Epidemiology*, 87, 70–77. https://doi.org/10.1016/j.jclinepi.2017.04.003

Siegfried, T. (2014) Scientists' grasp of confidence intervals doesn't inspire confidence, *ScienceNews*. http://www.sciencenews.org/blog/context/scientists-grasp-confidence-intervals-doesnt-inspire-confidence (accessed 21 April 2022).

6

AN INTRODUCTION TO DATA STORIES

An overview of this chapter

Setting the scene for why we need data is an important first step, whether it's to answer a research question, to create new knowledge, or to produce an engaging and convincing data story. This chapter will touch on the ontology and epistemology of knowledge, the scientific method, and what makes good research questions. There will also be a discussion on the value of research and good storytelling, including who cares about it.

Resources to support this chapter

This chapter draws on a range of sources, to help show you what makes a good data story. You can build your skills by trying the suggested activities for constructing clear and concise research hypotheses, as well as commenting on various data displays, in terms of their clarity and usefulness.

Resource	Date accessed	Location
The prevalence and correlates of accurate singing, *Journal of Research in Music Education*, 2020	20 October 2021	https://journals.sagepub.com/doi/abs/10.1177/0022429420951630
Can't sing? Do it more often, *ScienceBlog*, 2015	20 October 2021	https://scienceblog.com/76894/cant-sing-often/
Brexit by numbers, *Sky News*, 2020	04 March 2022	https://news.sky.com/story/better-for-brexit-how-uk-has-changed-since-leave-vote-11920143
SENSITIVE DATA: Jews of Lebanon, *Arab News*, 2022	04 March 2022	https://www.arabnews.com/JewsOfLebanon
Grammarly	26 July 2022	https://www.grammarly.com/

(Continued)

(Continued)

Resource	Date accessed	Location
Rishi Sunak wins race to become the UK's new prime minister, *Al Jazeera*, 2022	30 December 2022	https://www.aljazeera.com/news/2022/10/24/rishi-sunak-becomes-uks-new-prime-minister
What is an influencer? – Social media influencers defined [Updated 2022], *Influencer Marketing Hub*, 2022	30 July 2022	https://influencermarketinghub.com/what-is-an-influencer/
Cost of living: Almost half of adults finding it difficult to afford their bills – with numbers rising, *Sky News*, 2022	26 July 2022	https://news.sky.com/story/cost-of-living-almost-half-of-adults-finding-it-difficult-to-afford-their-bills-with-numbers-rising-12729637

The websites referred to in all the activities are provided as part of this book's online resources. You can find them at **https://study.sagepub.com/jonesstatsliteracy**.

6.1 Knowledge, reality and the scientific method

Knowledge exists in many forms and can be viewed as the basis of humanity's desire to develop expertise in any given subject. Over the ages, the quest for knowledge has led to many interpretations and perspectives as to the meaning of life, why we are here, as well as an exploration of phenomena and life forms that exist around us in the universe. Questions that are based on existence, reality and being are philosophically situated, often referred to as ontological in nature. Epistemological considerations refer to the philosophy of knowledge, which requires consideration of the relationship between knowledge, truth, belief, reason, evidence and reliability. Examples of epistemological questions are: What is knowledge? How is knowledge acquired? What do people know? How do we know what we know?

The approaches used to generate and analyse the data needed to reaffirm and create new knowledge are partly determined by the discipline in which it sits. For example, chemists, physicists and biologists often follow the scientific method to generate data to support or refute a hypothesis they have come up with to solve a problem or fill a gap in knowledge. For most of the science-based disciplines, scientific training often does not even consider epistemological and ontological perspectives, ardently following the scientific method without even a second glance. Scientists often follow a deductive approach to gain new knowledge and build on an existing evidence base to make a new contribution to the esoteric disciplinary groupings they belong to. Of course over time, new contributions can bring existing knowledge into question, and these need to be supported by hard evidence that has been repeated and investigated in depth, as well as being peer-reviewed by members of their esoteric group, to ascertain its reliability and validity.

The four steps of the scientific method are:

1   Observation – of some measurable aspect of the universe
2   Hypothesis – about a property of the universe, based on observation

3   Prediction – of something that should hold true if the hypothesis is correct
4   Experiment – to test the prediction

The steps can be repeated indefinitely, and any repetition will either support, refute or modify the existing theory.

Sociologists (and many other humanities-based researchers and academics), on the other hand, view the social world in many ways, and this can determine the approaches they take to generate data. For example, social scientists may refute the use of the scientific method to measure gender interactions in the classroom, arguing that qualitative approaches are more appropriate to create rich and detailed data, while quantitative approaches can be reductionist in order to narrow a focus. For them, observational data taken over time (e.g., longitudinally), could provide much more insightful information than using quantitative approaches (e.g., a class survey). The key methods used to generate data to help inform the creation of new knowledge are discussed in more detail in Chapter 8.

········Develop your skills! 6.1 ···· ···

The following questions will help you to further develop your skills in critical thinking, and get you to think about what knowledge is, as well as the scientific method.

Figure 6.1 Philosophical ideas
Source: Photo by cdd20 on Unsplash

(Continued)

1   How do you know what is beneath the ground? Is the moon there when you are not looking at it? How do you know if your first memory is real or made up? How do you question what is real and not real? Have a think and jot down a few notes.
2   Outline the four steps of the scientific method.

. . . .

Try to avoid:

1   Assuming one method or way of thinking about what we consider to be valid and legitimate ways of collecting data and doing research are better than others. They are all important and have advantages and disadvantages in different situations, applicable to different kinds of research.

6.2 What makes a good research question?

A research question can often be posed as a hypothesis (which we came across in Chapter 4), especially for individuals using the scientific method. This process often involves a deductive approach, which is concerned with 'developing a hypothesis (or hypotheses)' based on existing theory, and then designing a research strategy to test the hypothesis. Research questions should consider other people's work, and what others have done, to help build or refute existing knowledge and ideas. **Ethical considerations** should also be borne in mind when designing research questions, which can include the need to send research proposals to an ethics committee, who are trained to assess if the proposed research and research questions are ethically sound and have been designed to mitigate or reduced potential harm to participants or animals.

6.2.1 What is a hypothesis?

We have come across hypothesis testing in the previous chapter; this section will guide you to further build your skills in developing good research questions and associated hypotheses. A hypothesis is a tentative statement that proposes a possible explanation to some phenomenon or event. They are often used in deductive reasoning, as part of a scientific method. Note that hypotheses in this chapter are posed as statements, which can be useful to develop a hunch or idea, often based on existing knowledge and/or data. They can be further developed and converted into null and alternative hypotheses, as presented in Chapter 4.

A useful hypothesis is a testable statement which may include a prediction. Usually, a hypothesis is based upon a previous observation such as noticing that in the northern hemisphere in November, the colour of leaves changes.

How are hypotheses written? Here are some examples:

1   Educational attainment may be affected by ethnicity
2   Bacterial growth may be affected by temperature

3   Ultraviolet light may cause skin cancer
4   Viruses may cause serious disease in humans and animals

All the above are hypotheses because of the use of the word 'may'. The word 'may' suggest that one factor might be causing the change in another otherwise separate entity. However, the wording makes no suggestion as to how you might go about testing an observation. To write a scientifically testable hypothesis, we need to use a formalised hypothesis, for example:

> If ultraviolet (UV) light is related to skin cancer, then people with a high
> exposure to UV light will have a higher frequency of skin cancer.

The words 'if' and 'then' are used to link the two factors (UV light and skin cancer) together. The end statement is a tentative prediction as to what might be expected if a study was carried out.

········ Develop your skills! 6.2 ··· ···

The following questions will help you to develop your ability to write formal hypotheses.

Figure 6.2   Research questions
Source: Photo by Emily Morter on Unsplash

Write a formalised hypothesis for the following:

1   Bacterial growth may be affected by temperature
2   Chocolate causes spots

(Continued)

3   Educational attainment may be affected by ethnicity
4   An individual's lifespan may be affected by diet

···

········ Develop your skills! 6.3 ··· ···

The following questions will help you to further develop your ability to spot and interpret
research questions, presented in different formats.

Figure 6.3   Image of a choir in Nigeria

Source: Photo by John Onaeko on Unsplash

Using the website for the resource titled 'The prevalence and correlates of accurate singing' (https://
journals.sagepub.com/doi/abs/10.1177/0022429420951630), read the abstract to this journal article.
Answer the following questions linked to the resource:

1   The article looks at different factors that can influence singing accuracy. Who do you think would
     find this type of research interesting? Who would this be of value to?
2   Is the research question in the abstract clear?

Now look at this webpage resource, linked to the journal articles findings: 'Can't sing? Do it more often'
(https://scienceblog.com/76894/cant-sing-often/) and answer the following questions:

3   How does this article compare to the journal article, 'The prevalence and correlates of accurate
     singing'?
4   Are the research questions in this article clearer now? Describe them if you can.

···

Try to avoid:

- Constructing hypothesis that you cannot answer or find appropriate methods to answer.
- Using the word 'prove'. Proof of a relationship between variables, or causal relationship, can be difficult to state, which is why language like 'this suggests' or 'it is highly likely that' are safer and more appropriate.

6.3 What data have you got and where did they come from?

The democratisation of data storytelling in the West has meant that the barriers to telling stories with data have radically diminished in recent years. Why? Because of several related trends.

1   *More data*. 'Big data' has become a cliché, but it's undeniable that we are producing and storing more data than ever, across every field imaginable, including the environment, property, finance, transport, entertainment, law and order, sport, epidemiology, data analytics, bioinformatics – the list is endless.
2   *Digitisation of data*. Certain fields where data and information have come in the form of written text, old manuscripts, even artefacts, are increasingly becoming digitised and stored on software and hardware devices. Subjects like the arts, history and even archaeology are all benefiting from creating digital forms of data, which makes access and analysis a lot easier than in the past.
3   *Better access to good data*. This isn't true across the board, but data are generally becoming more accessible. Government agencies are becoming more adept at proactively releasing data sets for analysis and reuse. Agencies are motivated to help the private sector make business decisions, but such data is available for everyone to use.
4   *Better internet connections and IT equipment*. Increased access to good internet connections, as well as more and more people owning laptops, desktops, tablets and mobile phones, means audiences are bigger, demand is higher, and more people want to read digital content that includes data stories.

Greater access to data has many advantages and can help us as to stay informed as individuals. It can also provide benefits to companies, governments, and all sorts of organisations, in many ways. However, we do need to be confident in scrutinising data, and ask ourselves who cares about the data? Why was it collected? Who is it based on? There is always an agenda behind data collection, a reason for what's included and what's not included, what's asked and what's not asked. Western data has a history of favouring white males, to the detriment of the representativeness of the data. For example, what about data on females? Queer data? And what about data on ethnic minorities?

There are green shoots starting to appear, as we address data inequities in the West. For example, Caroline Perez has written an excellent book titled *Invisible Women: Exposing Data Bias in a World Designed for Men*, which provides an eye-opening analysis of the gender politics of knowledge and ignorance. With examples spanning from technology to natural disasters, this is a useful source to help reflect on the biases that exist in data collection. Kevin Guyan has also written an illuminating book titled *Queer Data*, defined as data relating to gender, sex, sexual orientation and trans identity/history. The author writes about how current data practices reflect an incomplete account of LGBTQ lives and attempts to explain how data biases are used to delegitimise the everyday experiences of queer people. Robert T. Teranishi and his colleagues have produced an eye-opening book, titled *Measuring Race: Why Disaggregating Data Matters for Addressing Educational Inequality*, which pulls together the expertise of scholars from a range of disciplines to explore the current state of racial heterogeneity, data practice, and educational inequality. The book looks specifically at the implications of these issues in relation to educational practices and inequalities.

There is always an agenda behind why certain data are collected, which is an important question we need to constantly ask ourselves, since so many important decisions are based on it. For example, when the newest iPhone model is released, are the dimensions more suited to men or women? When family credits are distributed to family households, who is the money given to? Which data are governments using to help inform these decisions? How can governments and local councils ensure that the money is used to feed everyone in the family household?

········· Develop your skills! 6.4 ··· ···

SENSITIVE DATA EXAMPLE: Race and religion

The following questions are linked to an interesting example, based on race and religion. It will help you to be able to evaluate a data story, as well as looking into why the data were collected to support the related information.

Figure 6.4   Image of Beirut, a city in Lebanon
Source: Photo by Piotr Chrobot on Unsplash

Scroll through the webpage titled 'Minority report: The Jews of Lebanon' (https://www.arabnews.com/JewsOfLebanon), paying attention to the data displays throughout. Answer the following questions:

1   Comment on the presentation of the data displayed on this webpage. Are the data displays easily understandable?
2   Who do you think collected the data presented in the story? Why do you think the data were collected?
3   This data story has been labelled 'sensitive' (using the definition in this book). Outline reasons why you think this story is potentially sensitive for some.
4   Can you think of any improvements to the webpage which might help with the flow of information presented?

· ·

· · · · · · **Develop your skills! 6.5** ·

The following questions will help you to further develop your skills in being able to critique a data story.

Figure 6.5   Image depicting Brexit, in London
Source: Photo by Fred Moon on Unsplash

Scroll through the webpage titled 'Brexit by numbers' (https://news.sky.com/story/better-for-brexit-how-uk-has-changed-since-leave-vote-11920143), paying attention to the data displays throughout. Answer the following questions:

1   Comment on the presentation of the data displayed on this webpage. Are the data displays easily understandable?
2   Who do you think collected the data presented in the story? Why do you think these data were collected?
3   Can you think of any improvements to the webpage which might help with the narrative? And the flow of information presented?

· ·

····· ··Develop your skills! 6.6··· ···

The following questions will help you to further develop your skills in being able to critique a data story, linked to the Grammarly writing skills application.

Figure 6.6 Writing skills
Source: Photo by Arron Burdon on Unsplash

You have come across this webpage in Chapter 2. Scroll through the webpage for the Grammarly writing skills application (https://www.grammarly.com/), paying attention to the data displays throughout. Answer the following questions:

1 Comment on the presentation of the information displayed on this webpage. Are the data displays easily understandable?
2 Is the webpage convincing? Does it make you want to download and use the application?
3 Can you think of any improvements to the webpage which might help with the narrative? And the flow of information presented?

····· ··Develop your skills! 6.7··· ···

The following questions will help you to further develop your skills in being able to evaluate a data story.

Figure 6.7 Image depicting Big Ben, London
Source: Photo by Adi Ulici on Unsplash

Scroll through the webpage titled 'Rishi Sunak wins race to become the UK's new prime minister' (https://www.aljazeera.com/news/2022/10/24/rishi-sunak-becomes-uks-new-prime-minister), paying attention to the data displays throughout. Answer the following questions:

1 Comment on the presentation of the data displayed on this webpage. Are the data displays easily understandable?
2 Who do you think collected the data presented in the story? Why do you think these data were collected?
3 Can you think of any improvements to the webpage which might help with the narrative? And the flow of information presented?

· ·

· · · · · · **Develop your skills! 6.8** ·

The following questions will help you to further develop your skills in being able to critique a data story, linked to the world of social media influencers.

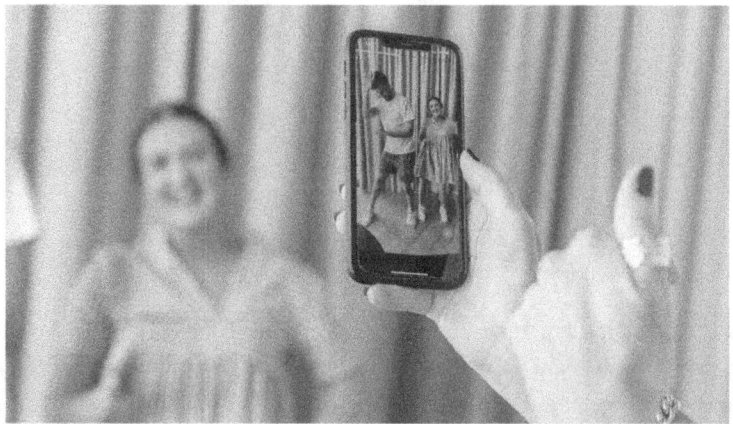

Figure 6.8 Image depicting social media influencer activity
Source: Photo by Social Cut on Unsplash

Scroll through the webpage titled 'What is an influencer? – Social media influencers defined [updated 2022]' (https://influencermarketinghub.com/what-is-an-influencer/), paying attention to the data displays throughout. Answer the following questions:

1 Comment on the presentation of the data displayed on this webpage. Are the data displays easily understandable?
2 Who do you think collected the data presented in the story? Why do you think this data was collected?
3 Can you think of any improvements to the webpage which might help with the narrative? And the flow of information presented?

· ·

The following questions will help you to further develop your skills in being able to critique a data story, linked to an interesting political example.

Figure 6.9   Image depicting 'How the Tories have increased the cost of living' poster, illustrating the increase in the price of tea and sugar under a Tory government (Tory government of 1905 compared to Liberal government of 1895), produced for the Liberal Party, c.1905–1910

Source: Photo by LSE Library on Unsplash

Scroll through the webpage titled 'Cost of living: Almost half of adults finding it difficult to afford their bills – with numbers rising' (https://news.sky.com/story/cost-of-living-almost-half-of-adults-finding-it-difficult-to-afford-their-bills-with-numbers-rising-12729637), paying attention to the data displays throughout. Answer the following questions:

1   Comment on the presentation of the data displayed on this webpage. Are the data displays easily understandable?
2   Who do you think collected the data presented in the story? Why do you think these data were collected?
3   Can you think of any improvements to the webpage which might help with the narrative? And the flow of information presented?

Try to avoid:

1   Assuming all research, which can include the approaches taken, is truly objective and free from bias. Even the scientific method is biased in several ways.
2   Trusting all data you come across. Remember that *all* data has a hidden agenda, a reason why it was collected in the first place. Think about these reasons and ask yourself if the data you are seeing could be missing something useful to give you a fuller picture of the theme or topic of the data.

Key points to remember

1   Think carefully about how you view the world around you, especially what you consider to be legitimate ways of collecting data.
2   Ask yourself questions that push you to think about whether you trust the data you are observing, as well as paying attention to how they are embedded into a data story.
3   Think carefully about what makes a good research question. Explore different research questions you have come across and ask yourself if you think they are measurable and if they can be answered. You might need to also think about the appropriateness of the methods chosen to answer the related research questions.

Further reading to support this chapter

Guyan K. (2022). *Queer Data: Using Gender, Sex and Sexuality Data for Action*. Bloomsbury Publishing.
Perez C. (2019). *Invisible Women: Exposing Data Bias in a World Designed for Men*. Chatto and Windus.
Teranishi, R.T., Nguyen, B.M.D., Alcantar, C.M. and Currameng, E.R. (2020). *Measuring Race: Why Disaggregating Data Matters for Addressing Educational Inequality*. Teachers College Press.

References to support this chapter

Al Jazeera (2022) Rishi Sunak wins race to become the UK's new prime minister. http://www.aljazeera.com/news/2022/10/24/rishi-sunak-becomes-uks-new-prime-minister (accessed 30 December 2022).
Garcia, C.A. and Heffer, G. (2020) Brexit by numbers: For better or for worse? How divorce from EU is working out for UK so far, *Sky News*. news.sky.com/story/better-for-brexit-how-uk-has-changed-since-leave-vote-11920143 (accessed 4 March 2022).
Geyser, W. (2022) What is an influencer? – Social media influencers defined [updated 2022], *Influencer Marketing Hub*. https://influencermarketinghub.com/what-is-an-influencer/ (accessed 30 July 2022).
Harwood-Baynes, M. (2022) Cost of living: Almost half of adults finding it difficult to afford their bills – with numbers rising, *Sky News*. news.sky.com/story/cost-of-living-almost-half-of-adults-finding-it-difficult-to-afford-their-bills-with-numbers-rising-12729637 (accessed 26 July 2022).
Kossaify, E. and Ziedan, N (2020) Minority report: The Jews of Lebanon, *Arab News*. http://www.arabnews.com/JewsOfLebanon (accessed 4 March 2022)
Pfordresher, P.Q. and Demorest, S.M. (2020) The prevalence and correlates of accurate singing. *Journal of Research in Music Education*, 69 (1), 5–23. https://doi.org/10.1177/0022429420951630
ScienceBlog (2015) Can't sing? Do it more often. https://scienceblog.com/76894/cant-sing-often/ (accessed 20 October 2021).

7

MEDIA REPORTS AND SOCIAL MEDIA PLATFORMS

An overview of this chapter

The use of data in the media and social media platforms is ubiquitous. They often form the foundation for engaging (and sometimes misleading) news stories, as well as business and commercial advertisements. This chapter draws on a range of examples, from the media and social media, emphasising the use of language to present a certain point of view. This chapter will build on the skills you have developed in Chapters 2–6, reinforcing the essential statistical literacy and critical thinking skills covered in them. 'Develop your skills' tasks will build on the examples presented, to enable you to gain a depth of understanding as to how the media use language (and the potential biases they have) when communicating data.

Resources to support this chapter

This chapter draws on a range of sources, to help show you how different media and social media platforms use and communicate data. We use a range of real-world examples to show you why certain language is used to persuade the reader of the author's point of view or agenda. You can build your skills by trying the suggested activities to help build your critical thinking skills, in relation to the areas mentioned above.

Resource	Date accessed	Location
Social media as a language learning environment: A systematic review of the literature (2008–2019), *Computer Assisted Language Learning*, 2021	09 June 2022	https://www.tandfonline.com/doi/full/10.1080/09588221.2021.1883673?casa_token=dXL43YeJ_9sAAAAA%3ArJ3rGMt70XC-vTo_yTEfgIs6Y6dJ7PAWz2mG73-GIHPzBxU_o0wS0CFyjbtUtbwU7jtyPPw-zfri
Speaking the language of data is now more important than ever, *TSG*, 2020	09 June 2022	https://www.tsg.com/blog/big-data/speaking-language-data-now-more-important-ever

(Continued)

(Continued)

Resource	Date accessed	Location
SENSITIVE DATA: Covid-19 news and information: consumption and attitudes – previous results, Ofcom, 2021	09 June 2022	https://www.ofcom.org.uk/research-and-data/tv-radio-and-on-demand/news-media/coronavirus-news-consumption-attitudes-behaviour/previous-results
SENSITIVE DATA: Covid-19 news and information: consumption and attitudes – Week 1 of UK lockdown, Ofcom, 2020	09 June 2022	https://www.ofcom.org.uk/__data/assets/pdf_file/0031/193747/covid-19-news-consumption-week-one-findings.pdf
SENSITIVE DATA: Covid-19 news and information: consumption and attitudes – interactive data, Ofcom, 2021	09 June 2022	https://www.ofcom.org.uk/research-and-data/tv-radio-and-on-demand/news-media/coronavirus-news-consumption-attitudes-behaviour/interactive-data
SENSITIVE DATA: Fake news, BBC, 2022	09 June 2022	https://www.bbc.co.uk/news/topics/cjxv13v27dyt/fake-news
ASOS: Topshop, 2022	09 June 2022	https://www.asos.com/topshop/
Almost 50 shops a day disappear from High Streets, BBC, 2021	09 June 2022	https://www.bbc.co.uk/news/business-58433461
There's science behind your inexplicably close relationship with your cat, Quartz, 2014	09 June 2022	https://qz.com/304236/theres-science-behind-your-inexplicably-close-relationship-with-your-cat/.

The websites referred to in all the activities are provided as part of this book's online resources. You can find them at **https://study.sagepub.com/jonesstatsliteracy**.

7.1 Am I data literate? (The need for statistical literacy)

Telling stories with data often requires several key skills, a certain level of statistical literacy being one of them. Statistical literacy, sometime called data literacy, can be defined as the ability to understand and reason with statistics and data. The literacy elements require the use of good writing skills, being able to craft a coherent and concise narrative, intertwined with data and numbers. Many of the skills you have started to develop already in this book have been guiding you towards improving your statistical literacy skills.

It is important to bear in mind that skills related to statistical literacy are not easy to develop, and it is not always straightforward to bring them together into a coherent and complimentary set of statistical tools. However, with perseverance and practice, it is hoped that you will be able to see how valuable they really are, and that you will find them useful. Try not to worry that you need to have mastered all the skills covered in this book, straight after you have covered the relevant content. Learning can be a different journey for many people, and in the case of statistical literacy, it takes time to master all the relevant skills. So be patient!

7.2 How the media use language and data

The use of data in news stories has a long tradition which dates back centuries. However, with the advent of new computer technologies, and the introduction of the World Wide Web, data have never been so readily accessible. We live in a world full of data, presented on 24-hour news television channels, in magazines, social media applications and email messages. The presentation of data needs to be eye-catching, understandable, and relevant to the end user. Often the data we see, on social media platforms and online news channels, are presented in easy-to-read graphs, with percentages being a popular form of data communication.

The content we see in news stories and social media platforms is often short and needs to be punchy (increasingly, reading times are stated at the beginning of news stories to help you to decide if you want to invest the time into reading them). As we become increasingly bombarded with data and news stories, our concentration spans have noticeably decreased in recent decades. These considerations are having an increasing impact on the size and types of data displays authors construct.

When we delve deeper into news stories on social media platforms and news channels, we quickly begin to see certain patterns emerging, as well as interesting insights into the types of stories that become popular. News channels, and news reports on other TV channels. often communicate stories that have negative connotations, with positive, feel-good stories usually in the minority. If we pause for a moment and reflect on these observations, we could ask ourselves why this might be the case. Is it because negativity bias in news stories is leveraged by the media to increase profits? Bad news gets more attention, more clicks, and leads to more revenue for publications. The demands for gloom and doom stories are likely explained by what psychologists have long recognised as our 'negativity bias' in that we pay more attention to and better remember negative experiences.

········Develop Your Skills! 7.1 ··

The following questions will help you to develop your skills in looking at data displays, and evaluating the language used within them. This will also help to develop your statistical literacy skills.

Figure 7.1   Learning and social media
Source: Photo by Rodion Kutsaiev on Unsplash

(Continued)

Read the abstract of the article titled 'Social media as a language learning environment: a systematic review of the literature (2008–2019)' (at https://www.tandfonline.com/doi/full/10.1080/09588221.2021.1883673), then answer the following questions:

1 Is there anything that surprises you about the abstract of this article?
2 Think of a learning experience from school or university where social media was used as part of the associated activity. Describe the activity, and the impact it had on your learning.
3 If you cannot remember any learning experiences in Question 2, think of a way you could use social media to help you better understand statistical literacy. Make some notes on this and try it out on a friend or family member. For example, you could ask them to record how many times in a day an advert pops up on their Facebook account. You could then ask them to look on different days of the week to see if there is a difference.
4 How was the activity in Question 3 received by your friend or family member? Did they like it? Did it help them better understand a statistical concept or idea?

· ·

· · · · · · · **Develop your skills! 7.2** ·

The following questions will help you to further develop your skills in looking at data displays, and evaluating the language used within them. This will also help to develop your statistical literacy skills.

Figure 7.2 Journalism
Source: Photo by Marek Pospíšil on Unsplash

Read the article titled 'Speaking the language of data is now more important than ever' (at https://www.tsg.com/insights/speaking-the-language-of-data-is-now-more-important-than-ever/), then answer the following questions:

1   Do you agree with the article's points? Explain your answer.
2   Do you think the article makes a strong case for data literacy? Explain your answer.

The article finishes by stating these four points:

> So how do we meet this challenge head on? It's simpler than you think – and you don't
> need to be a data scientist to do it!
>
> 1   Start by being *curious*. When you read an article or see some data, find out where it is from.
>   Cross-reference it with another source. Only when you are sure of its validity do you spend
>   time thinking about and perhaps sharing it.
> 2   Ditch the news sites and outlets that you *know* to be sensationalist or false. I saw a news article
>   the other day about the army being deployed around the M25. The accompanying picture in the
>   article was of a *Russian* tank – ditched.
> 3   Cleanse your social media timelines. Get rid of fake news. This will hurt, but I *promise* you it is
>   worth it. Your Facebook/Twitter feed will be game-changing.
> 4   Try your own. There is plenty of data out there and some simple tools to do it. Qlik Sense and
>   Power BI are particularly good for beginners wanting to skill up. You don't need to be a data
>   scientist to do this, you just need a curious mindset. There are so many charts and visuals out there
>   it's mind-boggling, so I decided to build my own using Qlik Sense and data from a trusted source.

3   Are there any problems you see with these points? How practical is it to implement what the
    article suggests here?

· ·

· · · · · · **Develop your skills! 7.3** ·

SENSITIVE DATA EXAMPLE: Covid-19

This task will help you to further develop your skills in looking at data displays, from different
sources based on Covid-19. It will also help you in being able to evaluate the language used
within them, which will also help to develop your statistical literacy skills.

Figure 7.3   Spread of Covid-19
Source: Photo by Yoav Aziz on Unsplash

(Continued)

Read the Ofcom article titled 'Covid-19 news and information: Consumption and attitudes – previous results', 2021 (at https://www.ofcom.org.uk/research-and-data/tv-radio-and-on-demand/news-media/coronavirus-news-consumption-attitudes-behaviour/previous-results), then answer the following questions:

1   Describe the arrangement of data (such as percentages) in the article. Are they presented effectively?
2   Part of the article reports misinformation. How do you think people are able to spot false or misleading information?

Now read the Ofcom report titled 'Covid-19 news and information: Consumption and attitudes – Week 1 of UK lockdown', 2020 (at https://www.ofcom.org.uk/__data/assets/pdf_file/0031/193747/covid-19-news-consumption-week-one-findings.pdf), then answer the following questions:

3   The report states on page 1 that 'the survey findings are representative of the views and habits of the 87% of the UK population that are online'. Is this a large enough sample to be representative of the UK population that are online?
4   From the key findings on page 1, why do you think the BBC was the most used service for people to gather information on Covid-19 online?
5   On page 3, the report states the following: 'Nearly half of respondents (46%) say they have come across false or misleading information about Covid-19 in the last week. Those aged 18–24 are more likely to say this (58%) compared to those aged 65+ (33%).' Why do you think younger people were more likely to say that they came across misleading information versus older people?
6   Look at the stacked bar graph on page 4. Very small percentages of people who filled in the survey reported that in week one they were following official guidance not very closely or not at all closely. Are there any types of non-sampling biases that could partly explain these small percentages?

Figure 7.4   Two people with face masks saying 'hi' and not using a handshake, to help reduce the spread of Covid-19

Source: Photo by Maxime on Unsplash

Finally, look at the dynamic data display on the webpage titled 'Covid-19 news and information: Consumption and attitudes – interactive data', 2021 (at https://www.ofcom.org.uk/research-and-data/tv-radio-and-on-demand/news-media/coronavirus-news-consumption-attitudes-behaviour/interactive-data), then answer the following questions:

7   Click on one of the coloured tiles in the middle of the webpage. Describe the data display behind the coloured tile you have clicked on.

8   Click on the red tile with the following text: 'Three in ten (32%) were "trying to avoid news about coronavirus" in week 76'. Now click on Q.1a: 'Typically, how often would you say you use social media'. Which response gave the lowest percentage of respondents, across the weeks listed on the graph?

....

Try to avoid:

1   Taking data displays at face value. Ensure you fully question the origins of a news story or social media platform and associated content. Ask if there might be an agenda behind the data story or information that is being presented.

2   Forgetting about relevant content that you picked up in earlier chapters. Much of the content and associated skills you have picked up in the previous chapters will help you to assess the reliability of news stories and social media platform posts.

7.3 Fake news

Fake news is a term that we see increasingly on the news and on social media platforms; however, it can mean different things to different people. In its broadest sense, fake news can refer to news stories or social media posts that are false: the story itself is fabricated, with no verifiable facts, sources or quotes. Sometimes these stories may be propaganda that is intentionally designed to mislead the reader or may be designed as 'clickbait' written for economic incentives (the writer profits according to the number of people who click on the story). In recent years, fake news stories have proliferated via social media, in part because they are so easily and quickly shared online.

The world of fake news is much larger than just false news stories. Some stories may be partially true; however, they can also be littered with inaccuracies and other false statements. Verifiable facts or sources may also be scarce. Some stories may include basic verifiable facts but are written using language that is deliberately polarising or sensationalist, leaves out pertinent details or presents only one viewpoint. Fake news does serve several purposes, which can include propaganda, misinformation and disinformation.

Misinformation is false or inaccurate information that is mistakenly or inadvertently created or spread; the intent is not to deceive. **Disinformation** is false information that is deliberately created and spread to influence public opinion or obscure the truth. The creation of misinformation and disinformation can be directly related to who the author (authors) is (are), which can help is to understand the reasons for their creation. Reasons for their creation can include:

1   Someone wanting to make money, regardless of the content of the article
2   Politicians or celebrities who have a hidden or covert agenda
3   People who want to either make a point or entertain you, or both
4   Journalists who are under sustained pressure to get the next big news story

The technological ease of copying, pasting, clicking, and sharing content online has helped these types of articles to disseminate more widely, and in some cases become viral. Articles can be designed to provoke an emotional response and placed on certain sites ('seeded') to entice readers into sharing them widely. Fake news articles can also be generated and disseminated by bots, which are computer algorithms that are designed to act like people sharing information but can do so quickly and automatically.

········ **Develop your skills! 7.4** ··

SENSITIVE DATA EXAMPLE: Fake news

This task will help you to further develop your skills in looking at data displays concerning fake news. It will also help you in being able to evaluate the language used within them, which will also help to develop your statistical literacy skills.

Figure 7.5   Fake news
Source: Photo by Markus Winkler on Unsplash

Look at the BBC 'Fake News' webpage (at https://www.bbc.co.uk/news/topics/cjxv13v27dyt), then answer the following questions:

1   Click on one of the stories on the BBC Fake News site and read the article.
2   Make notes on the news story. Is it eye-catching? Does it take long to read? How is the data
    displayed in news story?
3   What is it that the news story is tackling, related to fake news?
4   Are there any features from the news story you have selected that make you trust what you
    have read? Describe them.

· ·

· · · · · · **Develop your skills! 7.5** ·

This task will help you further develop your skills when looking at data displays. It will also
help you in being able to evaluate the language used within them, which will develop your
statistical literacy skills.

Figure 7.6   Shopping
Source: Photo by Jazael Melgoza on Unsplash

Review the webpage 'ASOS: Topshop', 2021 (at https://www.asos.com/topshop/), then answer the following questions:

1   Describe the layout of this webpage. Is it engaging? Make notes on how appealing the
    websites layout.

Now read the article titled 'Almost 50 shops a day disappear from High Streets', 2021 (at https://www.bbc.
co.uk/news/business-58433461), then answer the following questions:

2   After reading the news story, comment on the data displays and graphs you can see. Are they
    presented well and in an order that is easy to follow?
3   Do you think this news story is well balanced? Is it giving you the full picture? Think about the
    webpage you looked at when answering Question 1. Does ASOS have a physical store on most
    high streets in the UK?

· ·

···· Develop your skills! 7.6 ··· ···

This task is all about cats and their relationships with humans. The questions posed will help you to evaluate the data story presented, as well as paying attention to how the numbers are used in this specific example.

Figure 7.7   Cats and humans
Source: Photo by Chewy on Unsplash

Read the article titled 'There's science behind your inexplicably close relationship with your cat', 2014 (at https://qz.com/304236/theres-science-behind-your-inexplicably-close relationship-with-your-cat/), then answer the following questions:

1   Do you agree with the author's perspectives in this article?
2   How well does the author use facts and figures? Is the content referenced or backed up by others?
3   Would you consider this article to be fake news? Explain your reasoning.

· ·

Try to avoid:

1   Believing everything you read. Remember every article, news story or social media post was created by someone, who may have an agenda or reason for creating the story.
2   Thinking everything is fake news! Trusted sites (the BBC, NHS, *New Scientist*, and many others) will have lots of news stories that are reliable and include facts and usually good data displays. Of course the authors who create these stories will still have an agenda (or point of view they are trying to get across or persuade you to agree with them), but a good article should also present you with a balance of facts and perspectives, so that you can make up your own mind.

7.4 Framework for critically evaluating media stories

This chapter has presented you with various content and tasks to help build your skills in critically evaluating media stories. These skills are integral to the development of your statistical literacy abilities. Use the points below to further aid you in your quest to be able to assess the validity and reliability of media stories:

1   What is the topic of the news story or article? For example, is it to do with gender diversity? Or religion?
2   Who is the author? Are they a politician trying to get you to vote for their party? Or are they celebrity looking for more likes on a social media platform? The affiliation of an author can give you insights into why they have written a media story and what viewpoint they are trying to advance.
3   Why do you think the topic has been chosen? Try to think about what it can tell us about the world we live in. Is it interesting? What issue is it tackling or getting you to think about?
4   What part of the world is the story related to and is it time-sensitive? Is there something happening locally or further afield (e.g., an election campaign in the USA, or a war that has broken out on another continent) that could explain the content of a story? Sometimes the link between an event that is happening and a news story may not be so obvious at first.
5   When looking at the data in a media story, think about why they were collected, and how they are being used in the story. Look at the way the data are displayed and whether they make sense. Does the story follow a logical order, does it flow well? Are graphs explained well, are the axes labelled, are acronyms fully expanded or are they easy to interpret? These questions help to give us an idea of the statistical literacy elements of a media story, and how well they have been executed.
6   Finally, think about the intended audience of the report or story. This can really help you to understand the language a piece of work uses to deliver its key messages. It can also help you to unearth the reasons why the authors have taken a certain angle, and whether you think they have a case, that is, how persuasive the piece of work is.

Key points to remember

1   Ensure you can integrate and apply all the skills and knowledge you have nurtured and built upon in the previous chapters to this one (e.g., levels of measurement, how to write good sentences, using data displays, plots, and graphs). This will enable you to better evaluate the quality of media stories and content you see in news stories and on social media platforms.
2   Think about the language being used in media stories, who are the intended audience, how this frames the way the article or news story is presented.
3   Use the six points in Section 7.4 to help you to critically evaluate media stories, to enable you to decide if they are reliable and trustworthy.
4   Be patient and continue to practise and reflect upon the statistical literacy skills you are developing as you progress through the book.

References to support this chapter

ASOS (2022) ASOS: Topshop. https://www.asos.com/topshop/ (accessed 9 June 2022).

Barrot, J.S. (2021) Social media as a language learning environment: A systematic review of the literature (2008-2019). *Computer Assisted Language Learning*, 35 (9), 1–29. https://doi.org/10.1080/09588221.2021.1883673

BBC (2022) Fake news. https://www.bbc.co.uk/news/topics/cjxv13v27dyt (accessed 9 June 2022).

Guilford, G. (2014) There's science behind your inexplicably close relationship with your cat. *Quartz*. https://qz.com/304236/theres-science-behind-your-inexplicably-close-relationship-with-your-cat (accessed 9 June 2022).

Ofcom (2020) Covid-19 news and information: Consumption and attitudes – Week 1 of UK lockdown. https://www.ofcom.org.uk/__data/assets/pdf_file/0031/193747/covid-19-news-consumption-week-one-findings.pdf (accessed 9 June 2022).

Ofcom (2021) Covid-19 news and information: Consumption and attitudes – previous results. https://www.ofcom.org.uk/research-and-data/tv-radio-and-on-demand/news-media/coronavirus-news-consumption-attitudes-behaviour/previous-results (accessed 9 June 2022).

Ofcom (2021) Covid-19 news and information: Consumption and attitudes – interactive data. https://www.ofcom.org.uk/research-and-data/tv-radio-and-on-demand/news-media/coronavirus-news-consumption-attitudes-behaviour/interactive-data (accessed 9 June 2022).

Simpson, E. (2021) Almost 50 shops a day disappear from High Streets, *BBC*. www.bbc.co.uk/news/business-58433461 (accessed 9 June 2022).

Wannop, S. (2020) Speaking the language of data is now more important than ever, *Technology Services Group*. www.tsg.com/insights/speaking-the-language-of-data-is-now-more-important-than-ever/ (accessed 9 June 2022).

8

EXPERIMENTS AND OBSERVATIONAL METHODS IN RESEARCH

An overview of this chapter

Data are all around us! But where did they come from? How were they generated? Why were they generated? Can we trust them? What do we want to say about them? The next part of our data journey will help to answer these important questions and will require us to think about how we view the world around us, and how this determines our views on what we are willing to accept as reliable and valid data. The methods used to generate and analyse data should often be selected after the research questions have been identified. This chapter explores some of these concepts, and briefly touches on the philosophy of knowledge.

Resources to support this chapter

This chapter draws on a range of sources to help show you why different methods are used to collect data, and what conclusions you can draw from data collected using these different methods. In this chapter, we use a range of real-world examples to show you what good and bad sentences look like. You can build your skills by trying the suggested activities for evaluating different research methods used to collect data, as well as interpreting the associated research findings.

Resource	Date accessed	Location
When choice is demotivating: Can one desire too much of a good thing?, *Journal of Personality and Social Psychology*, 2000	14 August 2021	https://faculty.washington.edu/ jdb/345/345%20Articles/Iyengar%20 %26%20Lepper%20(2000).pdf

(Continued)

(Continued)

Resource	Date accessed	Location
An observational study on how situational factors influence media multitasking with TV: The role of genres, dayparts, and social viewing, *Media Psychology*, 2014	14 August 2021	https://www.tandfonline.com/doi/full/10.1080/15213269.2013.872038
Eating habits in the population of the Aeolian Islands: An observational study, *Public Health Nutrition*, 2018	15 August 2021	https://www.cambridge.org/core/journals/public-health-nutrition/article/eating-habits-in-the-population-of-the-aeolian-islands-an-observational-study/1A28F1A49CBCC8F502D15CC30D5948FD

The websites referred to in all the activities are provided as part of this book's online resources. You can find them at **https://study.sagepub.com/jonesstatsliteracy**.

8.1 Knowledge, validity and reliability

Knowledge exists in many forms and can be viewed as the basis of humanity's desire to develop expertise in a variety of subjects. Traditionally (especially in Western society), the acquisition of knowledge came from reading and researching facts and figures, which were contained in books often stored in libraries (Figure 8.1). With the World Wide Web becoming universally adopted in the 1990s, this is now the go-to place to search for information and data in all its forms.

Figure 8.1 Libraries have traditionally been the place to look for facts and figures
Source: Photo by Luke Tanis on Unsplash

The approaches used to generate and analyse the data needed to reaffirm and create new knowledge are partly determined by the discipline in which it sits. For example, chemists and biologists often follow the scientific method to generate data to support or refute a hypothesis they may have developed, to solve a problem or fill a gap in knowledge. They often choose experimental methods, whereby they control certain conditions and manipulate others to generate quantitative data. In using this approach, these scientists can replicate experiments to help improve the reliability of their results. In addition, if they can see similar patterns over time, scientists may be able to generalise these results externally (called external validity).

Sociologists, on the other hand, view the social world in many ways, which can determine the approaches they take to generate data. For example, social scientists may prefer to use other approaches to measure gender interactions in the classroom, arguing that qualitative approaches are more appropriate to create rich and detailed data. For them, qualitative approaches (which could include ethnography, a method that involves observing people in their own environment to understand their experiences, perspectives, and everyday practices) taken over time (e.g., longitudinally), could provide much more insightful information than using quantitative approaches (e.g., a class survey).

When we need to evaluate the claims made by the results or outputs from a study, as well as the statistical literacy proficiency of the associated reports, two key areas need to be reflected upon:

1   **Reliability** explains the degree to which a research instrument (e.g., a thermometer measuring temperature) measures a given variable consistently every time it is used under the same conditions with the same subjects. Reliability usually refers to data and not necessarily to measurement instruments. From different perspectives or approaches, researchers can evaluate the extent to which their instruments provide reliable data.

2   **Validity** refers to the accuracy of research data. A researcher's data can be said to be valid if the results of the study measurement process are accurate. That is, a measurement instrument is valid to the degree that it measures what it is supposed to measure. There are different types of validity. In an experiment, internal validity refers to whether there is a causal relationship between the variable being changed (called the independent or **explanatory variable**) and the variable being measured (called the dependent or **response variable**). In an observational study, internal validity refers to the strength of the findings from the study, that is, whether there is a causal relationship between factors that are being investigated, in relation to what is being measured. External validity refers to how well one can generalise research results to other settings, programmes, persons, places, etc. This can be applicable to both experimental and observa-tional studies.

These concepts can be difficult to understand at first, but they will become easier to understand as you work through this chapter and delve deeper into experimental and observational studies.

Try to avoid:

1   Assuming one type of method is better than another. A good researcher will choose methods that are the most appropriate to answer the research questions they have generated.
2   Thinking that one scientific discipline or subject area is better than another. All disciplines have their own strengths and weaknesses and can look at the world through a multitude of lenses.

········ **Develop your skills! 8.1** ··· ··

For this task, make notes on the terms covered in this section, using your own words where possible (i.e., paraphrase). When you have done this, read out what you have written to a friend, family member or pet (pets are likely to be the most willing, but they are the least likely to give you feedback!). If you are explaining the terms in this section, using your own words, to willing friends and family members, ask them if they understand what you have said. Tasks like should be useful, helping you to understand terms in this section.

Figure 8.2   Talking to your pet!
Source: Photo by Bruno Cervera on Unsplash

··

8.2 Experimental methods

An **experiment** involves a researcher changing or manipulating certain conditions, to see what effect this has on a response variable. In an experiment, the researcher determines which groups receive a treatment. The inclusion of a control group, who do not receive the treatment or intervention, can help in making useful comparisons between groups. The main aim of an experiment is to establish a causal relationship, which can help to ascertain whether a change in one variable leads to and causes a change in the response variable. In an experiment, there is often a random allocation of treatments to groups, to try to mitigate any bias in the data. Some studies include a blinding process, which can include both researchers and participants not knowing which group received the treatment. We call this a double-blinding process. If only the participants do not know which group received the treatment, we call this a single blinding process. This can help to reduce bias when the results are interpreted.

· · · · · · SENSITIVE DATA EXAMPLE: Smoking ·

Researchers in a group interested in childhood development of smoking behaviours have developed an intervention to help support children of parents who smoke, to try to prevent the children from smoking. In their study, they recruit a group of families, in which one of the parents has identified as being a heavy smoker. The researchers then randomly allocate families into two different groups. Those in group 1 receive a 20-week course to help educate parents and prevent their children from developing smoking behaviours, while those in group 2 receive no training over the 16 weeks (control group). All families are followed up over a 3-year period, with the children being asked to fill in a survey measuring whether they have developed smoking behaviours.

Figure 8.3   Smoking among teenagers
Source: Photo by Andrea Dibitonto on Unsplash

(Continued)

In this study the features which make it an experiment, are:

- The random allocation of treatment to the family groups
- The implementation of an intervention, which is the treatment (or explanatory variable)

If, after 3 years, the researchers find that the children in intervention groups show little or no signs of developing smoking behaviours (compared to the control group), then they might conclude that the intervention caused this change in behaviour. In addition, if the control group did show a marked increase in their children developing smoking behaviours, the researchers could be able to claim that their intervention helped to prevent this from happening in the intervention group.

. . . .

There are two important points to mention with this example. First, ethical considerations are very important, particularly with this type of research. Research (including observational studies) often requires the researchers to apply for ethical clearance from an ethics board. This ensures that the subjects involved in the study (which can include animals) are protected against any potential physical or psychological harm, and the appropriate safeguards are put in place if any harm does occur.

Second, experiments rely on the researcher controlling certain variables and manipulating others to see if there are any causal relationships present. In research like the above example, controlling certain variables can be difficult, if not impossible. We call these **confounding variables**. For example, in the above experiment, the researchers did not control for the age of participants (having been unable to select families that have similar ages to each other). This could have led to the inclusion of families with parents of vastly different ages. You could then argue that this variable (different ages of parents) could explain differences seen in children's smoking behaviours after the 3-year follow-up. For example, older parents may have been stricter and clung to traditional values, so their children showed lower levels of developing smoking behaviours, while younger parents may have been less strict. These issues will directly influence the internal and external validity of the experiment. However, the process of randomly allocating the treatment to families in this example could help to mitigate any confounding variables identified.

A commonly used experimental approach is a **randomised control trial** (RCT). RCTs involve the random allocation of people (or units) to receive one of several interventions. These can be clinical, sociologically based, or psychological interventions. One of these interventions is the standard of comparison or control. The control may involve a placebo that should not affect the response or dependant variable, or no intervention at all. A placebo (i.e., a 'sugar pill') should not have any causal effect on the dependant (or response) variable. Someone who takes part in an RCT is called a participant or subject. RCTs seek to measure and compare the outcomes after the relevant participants or groups receive the interventions.

Try to avoid:

1   Assuming that experiments are the best way to answer research questions and collect data.
2   Ignoring the ethical dimensions to experimental research, especially in the fields of psychology and sociology.

· · · · · · Develop your skills! 8.2 · · ·

The abstract below is taken from a well-known experiment in the field of psychology, which looked at personal choice and whether it is better to have more or less choice in helping humans make decisions. The study was undertaken by Sheena Lyenger and Mark Lepper, in the United States, and was published in 2000 in the journal *Personality Processes and Individual Differences*. Read the excerpt below, and answer the questions that follow:

Figure 8.4   Selection of jams
Source: Photo by Esther Pervis on Unsplash

When choice is demotivating: Can one desire too much of a good thing?

Current psychological theory and research affirm the positive affective and motivational consequences of having personal choice. These findings have led to the popular notion that the more choice, the better – that the human ability to manage, and the human desire for, choice is unlimited. Findings from 3 experimental studies starkly challenge this implicit assumption that having more choices is necessarily more intrinsically motivating than having fewer. These experiments, which were conducted in both field and laboratory settings, show that people are more likely to purchase gourmet jams or chocolates when offered a limited array of 6 choices rather than a more extensive

(Continued)

array of 24 or 30 choices. Moreover, participants reported greater subsequent satisfaction with their selections when their original set of options had been limited. Implications for future research are discussed.

1   Why do you think they performed three experiments on this topic?
2   What are the response and explanatory variables in this experiment?
3   Why do you think the experiments were conducted in the laboratory and in the field (i.e., in the real world)?
4   Do the results surprise you? Explain your answer.
5   From your own observations and experiences of being presented with choices, make a list of reasons to both support and refute the claims made in this paper.

. . . .

8.3 Observational methods

Observational studies often involve a researcher observing groups of people, animals or objects of interest, at a single point in time or over a longer period. These studies can include the collection of data using qualitative (e.g., group interviews) or quantitative approaches (e.g., surveys), or mixed methods (possibly a combination of qualitative and quantitative methods). The researcher might be involved with measuring changes to a response variable over time, looking for potential causal relationships, or they could be interested in collecting data over time to look for common themes.

Observational studies can be cross-sectional, providing a snapshot of data at a point in time on a research subject of interest, or longitudinal, following the research subjects over time, which can include data collection at multiple time points. Cohort studies, such as the Caerphilly Cohort Study in Wales and the Dunedin Study in New Zealand, are examples of longitudinal studies. Case–control studies can also be longitudinal; these compare two groups of people – those with the disease or condition under study (cases), and a very similar group of people who do not have the disease or condition (controls). These types of study are often used by medical researchers, so that they can look at the medical and lifestyle histories of the people in each group to learn what factors may be associated with the disease or condition. For example, one group may have been exposed to a particular substance, whereas the other group was not.

An example of an observational study could include a social science researcher investigating homophobic abuse in the night-time economy. The researcher might even go undercover and pose as a nightclub patron themselves (consider the ethical considerations in this example) to gain insider research perspectives. Their data might include observational data of physical environments that come under the umbrella of the night-time economy or could include transcribed recordings from conversations with other people engaged with clubbing and pubbing. The researcher could then look for common themes, perhaps from common words used to describe an event, mentioned in the recordings. This type of research is an example of a qualitative approach. It could also be longitudinal, if the researcher decided to take snapshots of data across a longer section of time.

As with experiments, ethical considerations need to be explored, as well as clearance provided from an ethics committee.

Confounding variables are a problem with observational studies, and some would argue that they are more problematic than with experimental approaches since variables that could influence any causal relationships proposed are not being controlled. For these reasons, observational studies are useful for identifying possible causes of effects, but they cannot reliably establish causation.

Try to avoid:

1   Assuming observational studies are not as useful as experiments. In the social world, using an experimental approach, particularly with random allocation processes included, is not always practical or desirable.
2   Assuming observational studies are not useful in identifying the possibility of a causal relationship between variables of interest.

· · · · · ·**Develop your skills! 8.3**· ·

The abstract below is taken from a popular journal article, where the researchers undertook an observational study to investigate factors that influence media multitasking when people are watching TV. In this study, media multitasking is studied at a behavioural level and defined as 'multiple exposures to various media forms at a single point in time for the same media consumer'.

Figure 8.5   Examples of social media platforms
Source: Photo by Alexander Shatov on Unsplash

(Continued)

The study was led by Hilde Voorveld and Vijay Viswanathan, in the United States, and was published in 2014 in the journal *Media Psychology* (see https://www.tandfonline.com/doi/abs/10.1080/1521326 9.2013.872038). Read the abstract and answer the questions that follow:

An observational study on how situational factors influence media multitasking with TV: The role of genres, dayparts, and social viewing

This study responds to the need for research on individuals' media multitasking behaviour using observational data. Media multitasking can have a profound impact on media processing and effects. However, we have little knowledge on when people are likely to engage in media multitasking and, consequently, when these effects are likely to occur. This study examines how three important situational factors – television genres, dayparts, and social viewing – influence the amount of media multitasking. Granular observational data obtained by directly monitoring and recording media consumption behaviours of a large panel at 10-second intervals are used for the analysis. The study reveals that media multitasking with television is most prevalent when people watch sports or engage in channel surfing and less prevalent with commercials, news, and entertainment. Furthermore, the extent of media multitasking is greater in the morning and afternoon than in the evening, and greater when individuals watch television alone than in the presence of others. Daypart differences are larger for genres associated with incidental viewing (commercials and channel surfing) than for genres associated with intentional viewing (news, entertainment, and sports). Sports is the only genre that is associated with higher amounts of media multitasking when watching television with others.

1   What are the advantages of using an observational approach in this study?
2   Are the results surprising? Explain your answer.
3   What other factors could explain the results reported by the authors?

· ·

· · · · · · · **Develop your skills! 8.4** · · · ·

The abstract below is taken from an interesting journal article, where the researchers undertook an observational study to investigate the eating habits of a sample of individuals from the Aeolian Islands, north of Sicily, Italy. The study was led by Paolo La Spina and colleagues and was published in 2018 in the journal *Public Health Nutrition*. Read the abstract below, and answer the questions that follow:

Figure 8.6   Healthy eating
Source: Photo by Brook Lark on Unsplash

Eating habits in the population of the Aeolian Islands: An observational study

Objective: We conducted a study to describe food profile, health status and stroke risk factors in the population of the Aeolian Islands.

Design: Self-administrated questionnaires regarding eating habits, health status and stroke risk factors were obtained from a sample of the general Aeolian population. We analysed the difference from common healthy eating habits indicated by the Italian Institute of Nutrition.

Setting: Current evidence finds the Mediterranean diet is a protective factor for cardio- and cerebrovascular diseases. The Aeolian Islands are an interesting study setting because of their peculiarity in the epidemiology of cerebrovascular and neurodegenerative diseases.

Participants: Individuals (n [=] 586; age range 15–93 years; mean 52 (SD 18) years) living in the Aeolian Islands.

Results: We found low fish consumption in 13·3% and vitamin intake deficiency in 5·8% of participants. A marked excess of saturated fats was observed in 71·0% of participants. Sodium excess was reported almost in half of participants (49·0%). Eating habits were characterized by high consumption of fruits and vegetables, consistent use of olive oil and scanty use of cured meat. Health status as evaluated by the General Health Questionnaire was characterized by 'normal distress' level in the majority of participants.

Conclusions: Study findings show the eating habits and health status of the Aeolian people in an interesting setting of low incidence of cerebrovascular disease. This nutrition regimen has been proved to be protective against cerebrovascular disease. Nutrition is likely to contribute to the low incidence of stroke in this population.

(Continued)

1   Are the results surprising from the study? Explain your answer.
2   Are there any potential biases or non-sampling errors in the study? Explain your answers.

....

Key points to remember

1   Students are often taught that experiments are the best way to generate data (especially in the natural and physical sciences), helping to keep the researcher as objective as possible, and to minimise bias in data interpretation. It really is not as simple as this and depends more on the type of research you are undertaking, the research questions posed, and, more importantly, the way you view the social world.

2   Be careful how you interpret results from a study, paying close attention to the phrases you use to describe any potential causal relationships identified.

3   A good researcher always comments on the strengths and weaknesses of any research carried out. This could include issues linked to ethics, as well as the validity and reliability of the methods and results.

4   When reading a research paper or any kind of study, first look at the research questions posed. Good research will choose appropriate research methods that can best answer the research questions. If you must engage with research in your studies, make sure you follow this process and avoid jumping straight into selecting the method that seems the easiest or the one you are most comfortable with.

5   Remember there are always strengths and weaknesses to all research. Good papers often present recommendations for further research, which could be linked to the weaknesses of the paper. These provide opportunities to take their research further.

6   Do not fall into the trap of thinking that some research methods are better than others. All research methods have their strengths and weaknesses. Using mixed methods, involving quantitative and qualitative approaches, can provide you with powerful tools. However, this can also make it difficult to analyse and write up the research.

....

References to support this chapter

Iyengar, S.S., and Lepper. M.R. (2000) When choice is demotivating: Can one desire too much of a good thing? *Journal of Personality and Social Psychology*, 79(6), 995–1006. https://doi.org/10.1037//0022-3514.79.6.995

La Spina, P., Savica, R., Ciacciarelli, A., Cotroneo, M., Dell'Aera, C., Grillo, F. … Musolino, R. F. (2018) Eating habits in the population of the Aeolian Islands: An observational study. *Public Health Nutrition*, 22 (9), 1590–1596. https://doi.org/10.1017/S1368980018003397

Voorveld, H.A.M. and Viswanathan, V. (2014) An observational study on how situational factors influence media multitasking with TV: The role of genres, dayparts, and social viewing. *Media Psychology*, 18 (4), 499–526. https://doi.org/10.1080/15213269.2013.872038

....

9

HOW TO READ JOURNALS WITH QUANTITATIVE DATA

An overview of this chapter

Now that you have created a foundation in your statistical literacy skills, this book will begin to focus in on data displays and information that you are likely encounter in your studies, and potentially in other areas of your life. Emphasis will be placed on the communication of data analysis and results in journal articles, including the language used in inferential statistics (introduced in Chapter 3). Articles based on experiments and observational studies will be included, with associated questions embedded around them, linking directly to the next chapter.

Resources to support this chapter

This chapter draws on a range of sources, to help show you to evaluate journals with quantitative data. In this chapter, we use a range of real-world examples to show you what good and bad journals articles look like. You can build your skills by trying the suggested activities on evaluating the language used in journal articles with quantitative data, as well as encouraging you to think about improvements that can be made to the examples presented.

Resource	Date accessed	Location
When choice is demotivating: Can one desire too much of a good thing?, *Journal of Personality and Social Psychology*, 2000	14 August 2021	https://faculty.washington.edu/ jdb/345/345%20Articles/Iyengar%20 %26%20Lepper%20(2000).pdf

(Continued)

(Continued)

Resource	Date accessed	Location
An observational study on how situational factors influence media multitasking with TV: The role of genres, dayparts, and social viewing, *Media Psychology*, 2014	14 August 2021	https://www.tandfonline.com/doi/full/10.1080/15213269.2013.872038
Eating habits in the population of the Aeolian Islands: An observational study, *Public Health Nutrition*, 2018	15 August 2021	https://www.cambridge.org/core/journals/public-health-nutrition/article/eating-habits-in-the-population-of-the-aeolian-islands-an-observational-study/1A28F1A49CBCC8F502D15CC30D5948FD
How to present research, *Malaysian Family Physician*, 2006	03 August 2022	https://www.ncbi.nlm.nih.gov/pmc/articles/PMC4453119/

The websites referred to in all the activities are provided as part of this book's online resources. You can find them at **https://study.sagepub.com/jonesstatsliteracy**.

9.1 Why do I need to use journal articles?

Journal articles are often regarded as the gold standard in relation to the way new knowledge is created, verified and communicated. At first, reading journals can seem challenging – it is as if they are written in an incomprehensible alien language. The reason for this is that they are written for an expert audience in a formal style and follow a certain structure, which may be unfamiliar. Journal articles are considered the fount of knowledge; it is therefore vital that you become versed in being able to decipher their meaning, figure out if they include good methods and trustworthy results, and decide whether they are of value to you in your studies. It is common for your instructors or lecturers, who are usually well acquainted with key researchers and their works in the field of study you are involved with, to help guide you to important journal articles, and help you identify appropriate databases and journals that are trusted and present peer-reviewed papers. It is worth spending the time learning how to read them, though, as it is a skill that will come in handy whether you are writing up laboratory reports, literature reviews or longer pieces of work such as dissertations or theses. They will especially come in handy when you come to write your own journal articles as well!

····· Develop your skills! 9.1 ·····

Being able to translate disciplinary specific terminology and phrases is an important step towards being able to comprehend larger pieces of work, and especially journal articles. This task will help you to develop these skills.

Figure 9.1 Translation
Source: Photo by Mathias Reding on Unsplash

Look at the article we studied in Chapter 8, titled 'When choice is demotivating: Can one desire too much of a good thing?' (at https://faculty.washington.edu/jdb/345/345%20Articles/Iyengar%20%26%20Lepper%20(2000).pdf). Answer the questions related to this article, after the abstract (the abstract is also presented below):

Current psychological theory and research affirm the positive affective and motivational consequences of having personal choice. These findings have led to the popular notion that the more choice, the better – that the human ability to manage, and the human desire for, choice is unlimited. Findings from 3 experimental studies starkly challenge this implicit assumption that having more choices is necessarily more intrinsically motivating than having fewer. These experiments, which were conducted in both field and laboratory settings, show that people are more likely to purchase gourmet jams or chocolates or to undertake optional class essay assignments when offered a limited array of 6 choices rather than a more extensive array of 24 or 30 choices. Moreover, participants actually reported greater subsequent satisfaction with their selections and wrote better essays when their original set of options had been limited. Implications for future research are discussed.

(Continued)

1   Select five words you do not understand in the abstract, write or type them on a document,
    and define them. Use online resources to help you.
2   Summarise the abstract in two sentences.
3   Reflecting on your own personal experiences on this subject, does anything surprise you in
    their findings?
4   Is there anything confusing in the abstract? Why is it confusing?

. . . .

9.2 Typical structure

Articles are often located in academic journals, and they usually follow a standard format, divided into various sections. The name and order of these sections can vary from
journal to journal and across subject disciplines; however, understanding how an article
is constructed will help you quickly locate the information you need.

The following subheadings are typically found in most journal articles, although, as
already mentioned, these can vary:

Title. This give you some indication as to what the article is about.

Author and institution. This tells you who the authors are and where they work.
From this you can find out if any of them are known experts in the field and if
their institution is internationally recognised. This will involve a follow-up Google
search on the said authors and their respective affiliations (fancy term for where
they are professionally based – in a university, for example).

Abstract. This is a short summary, which will often include why the study was
conducted, the methods used and the key results or findings.

Introduction. This provides some context and background information about the
topic, including details of other research in this area, with references to other
relevant articles. The aims and goals of the study are also usually outlined, which
can include what questions it will answer and how the findings may contribute to
the understanding or advancement of knowledge in that subject. Overall the
introduction should fully justify why the research was conducted.

Methods. This section describes the approaches and procedures used. These
descriptions should be detailed enough to allow other researchers to replicate
them. Methodological considerations can also be included, which involve drawing
on theoretical considerations, which can lead to deeper discussions as to why
certain methods were chosen or not chosen, to answer any research questions
posed.

Results and Discussion. This section often involves a presentation of the findings
and the authors' interpretation of their significance and how they relate to existing
research in this field. Data are often presented as graphs, tables, plots, flow

diagrams, pictures, audio recordings, and video footage, and can also include quotes and transcripts.

Conclusion. Usually the final part of a journal article, the conclusion summarises the details from the Results and Discussion section that the authors wish to highlight, or feel are the most important. Some journals may incorporate the conclusions into a Discussion section.

References. This section lists details of the sources cited in the article, which can include links to data sets related to the article's investigation. References can be helpful for finding related literature and information on the subject or area of study.

Note that you do not need to read an article in the order in which it is presented and, depending on the type of information you need and the level of detail you require, you may not need to read every section. It can be useful to read the abstract and conclusion first and then look at any diagrams, structures and figures.

Try to avoid:

1   Assuming that all journal articles will have the same subheadings as the ones presented in this section. For example, some may have subheadings like 'Background', instead of 'Introduction'.

········ **·Develop your skills! 9.2 ···** ···

The role of media within Western society, and indeed elsewhere, has become pervasive. Its reaches can be felt in all walks of life, and its impacts on human behaviour and how we live our lives are irrevocable. The abstract below is taken from a popular journal article, where the researchers undertook an observational study to investigate factors that influence media multitasking when people are watching TV.

Figure 9.2 Typical television display
Source: Photo by Nicholas J Leclercq on Unsplash

(Continued)

Look at the article titled 'An observational study on how situational factors influence media multitasking with TV: The role of genres, dayparts, and social viewing' (at https://www.tandfonline.com/doi/full/10.1080/15213269.2013.872038; we have already come across this example in Chapter 8). Answer the questions related to this article, after the abstract (the abstract is also presented below):

This study responds to the need for research on individuals' media multitasking behavior using observational data. Media multitasking can have a profound impact on media processing and effects. However, we have little knowledge on when people are likely to engage in media multitasking and, consequently, when these effects are likely to occur. This study examines how three important situational factors – television genres, dayparts, and social viewing – influence the amount of media multitasking. Granular observational data obtained by directly monitoring and recording media consumption behaviors of a large panel at 10-second intervals are used for the analysis. The study reveals that media multitasking with television is most prevalent when people watch sports or engage in channel surfing and less prevalent with commercials, news, and entertainment. Furthermore, the extent of media multitasking is greater in the morning and afternoon than in the evening, and also greater when individuals watch television alone than in the presence of others. Daypart differences are larger for genres associated with incidental viewing (commercials and channel surfing) than for genres associated with intentional viewing (news, entertainment, and sports). Sports is the only genre that is associated with higher amounts of media multitasking when watching television with others.

1   Summarise the abstract into two sentences.
2   Comment on the layout of the journal article. Is it well presented? Is it clear? Do you understand what the study was about? Explain your answer.
3   Describe the layout of the introduction (it is called 'Background' in this article).
4   Are there any parts of the journal article you find unclear, or hard to understand? If yes, which parts and why?
5   How could the article be improved?

....

9.3 How do I evaluate a journal article? How do I know that it's any good?

The journal articles you read, which you may have found yourself or possibly were selected for you by your lecturers or instructors, are likely to have gone through the peer-review process before being published. This process involves reviewers, often researchers and experts in the relevant subject area, who have been tasked with evaluating the piece of work submitted for publication. They look at the methods and procedures used, to ascertain if the choices made are justified and appropriate. They also check the validity (potentially both internal and external) and reliability of the results, reflecting on the methods selected for the study (recall these terms from Chapter 8). They often look at the results in the context of the conclusions made, specifically scrutinising how the results have been processed and interpreted. They must also decide if the study is of value and worth disseminating to a wider audience, and if the article reflects the values and subject area of the journal they represent. Often the reviewers will suggest changes the authors must make if the article is to be accepted for publication. The aim of peer review is to maintain and improve the standard of published research. Articles in peer-reviewed journals are generally considered to be of high quality.

If an article has undergone the peer-review process, you might assume that you can simply accept the findings as trustworthy and of value. However, one of the skills of a good researcher is the ability to think critically and ask questions about the study. The process of critiquing articles will help you build your academic skills (which also links with the skills you picked up in Chapter 6). When you review articles, it also important to ensure you understand the subject-matter, and the subject area related to the article, to help you evaluate the methods and procedures used by the authors, and to help you to interpret the findings of the study.

The first steps in being able to critique a journal article will naturally involve reading the article in depth to critique it, so it is a good idea to skim through it at least once and look up any terms or processes you do not understand before you move on. You will find that over time, as you build up a bank of terms you need to translate, if you learn what they are and remember their meaning, the next time you come across them, you will start to pick up and understand larger sections of related journal articles a lot faster, which will also build your confidence and it will all begin to make a lot more sense! Try keeping a small book of terms or phrases that will help you better understand the subject-matter you are interested in or investigating.

Remember, the aim of critiquing is not necessarily to disprove the authors' conclusions or find errors in the results, although this can sometimes happen. The intention is to evaluate the study and the way it is presented. The impact of exciting and useful discoveries can sometimes be obscured in a poorly written article. While reading, consider if there are ways in which the article can be improved.

The following questions can help you ascertain the clarity of a journal article:

1   Did you find the information you expected in specific sections?
2   Were the results presented clearly and could they be easily understood?
3   How is the article presented? Is it laid out well? Easy to follow?
4   Were you convinced by the authors' interpretation of the results?

The rest of this section poses questions to help you think critically about the article as you read. If the article is well written and the study's findings are communicated well, you should be able to find the answers relatively easily in the suggested sections.

9.3.1 Why is this study important? What is the objective?

Read the Abstract and the Introduction to find out/look for:

1   An explanation of the purpose of the study; what it aims to achieve, the problem or issue it seeks to resolve or tackle
2   The significance and relationship of the study to the existing body of research in this subject, whether it is trying to investigate a gap in the literature or knowledge base of the subject area
3   The quality of the references used, and the use of relevant literature based on the topic of interest. Are there any key articles missing? Have the authors or author missed an important area out of the Introduction?

9.3.2 What was the overall plan of the author(s) to investigate the area of interest/subject-matter of the article?

Read the Methods section to find out/look for:

1   A description of the methods and techniques used
2   Have the authors provided sufficient detail so that others can replicate what they have undertaken, if applicable?
3   Justification for why these have been selected as opposed to other options

9.3.3 What were the authors' results and explanations for them?

Read the Results and Discussion find out/look for:

1   Relevant figures and tables which are clear and easily read, and appropriately labelled
2   Convincing and credible interpretation of the data

9.3.4 What are the main points from the Discussion and Conclusion?

Read the Discussion and Conclusion to find out/look for:

1   A clear summary of the findings supported by evidence from the Results section
2   Details of what the study accomplished

3   Whether the conclusions are justified by the results
4   Whether any biases or factors that could have influenced the data and related conclusions are fully explained and justified

9.3.5 What are the next steps – things that others should investigate further?

Read the Discussion and Conclusion to find out/look for:

1   Suggested further research, why it is required and what it should achieve
2   Details of any practical applications of the findings

· · · · · · Develop your skills! 9.3 ·

The abstract below is taken from an interesting journal article we saw in Chapter 8, where the researchers undertook an observational study to investigate the eating habits of a sample of individuals from the Aeolian Islands, north of Sicily, Italy.

Figure 9.3   Eating habits
Source: Photo by Spencer Davis on Unsplash

(Continued)

Look at the article titled 'Eating habits in the population of the Aeolian Islands: An observational study' (at https://www.cambridge.org/core/journals/public-health-nutrition/article/eating-habits-in-the-population-of-the-aeolian-islands-an-observational-study/1A28F1A49CBCC8F502D15CC30D5948FD#.) Answer the questions related to this article, after the abstract (the abstract is also presented below):

Objective: We conducted a study to describe food profile, health status and stroke risk factors in the population of the Aeolian Islands.

Design: Self-administrated questionnaires regarding eating habits, health status and stroke risk factors were obtained from a sample of the general Aeolian population. We analysed the difference from common healthy eating habits indicated by the Italian Institute of Nutrition.

Setting: Current evidence finds the Mediterranean diet is a protective factor for cardio- and cerebrovascular diseases. The Aeolian Islands are an interesting study setting because of their peculiarity in the epidemiology of cerebrovascular and neurodegenerative diseases.

Participants: Individuals (n [=] 586; age range 15–93 years; mean 52 (sd 18) years) living in the Aeolian Islands.

Results: We found low fish consumption in 13·3% and vitamin intake deficiency in 5·8% of participants. A marked excess of saturated fats was observed in 71·0% of participants. Sodium excess was reported almost in half of participants (49·0%). Eating habits were characterized by high consumption of fruits and vegetables, consistent use of olive oil and scanty use of cured meat. Health status as evaluated by the General Health Questionnaire was characterized by 'normal distress' level in the majority of participants.

Conclusions: Study findings show the eating habits and health status of the Aeolian people in an interesting setting of low incidence of cerebrovascular disease. This nutrition regimen has been proved to be protective against cerebrovascular disease. Nutrition is likely to contribute to the low incidence of stroke in this population.

1　Is the abstract presented well? Explain your answer.
2　Critically evaluate the journal article, using the prompts in Section 9.3. If you cannot access the whole journal article, critically evaluate the abstract.
3　Think about the causal claims the authors make in the conclusion. Are these justified? Explain your answer.

. . . .

9.4 Common data displays in journal articles

Journal articles, across a spectrum of disciplines, will invariably present data in some form. This could include data collected using quantitative or qualitative methods, or a mixture of the two. Using either approach could lead to the production of quantitative data. You have come across common data displays at several points in this book

(Chapter 3, and you will revisit this topic in Chapter 10), so now we will delve deeper into how to evaluate data displays, utilised by published journal articles, via a selection of tasks.

····· Develop your skills! 9.4 ··

The excerpt below is taken the same journal article we just looked at, where the researchers undertook an observational study to investigate the eating habits of a sample of individuals from the Aeolian Islands, north of Sicily, Italy.

Figure 9.4   Data displays
Source: Photo by Lukas Blazek on Unsplash

Answer the questions related to this article, and use the table and accompanying text below to help you, if you cannot access the full journal article:

Initially, 671 individuals were contacted for the enrolment. Eighty-five denied their consent or refused to participate. A total of 586 individuals (age range 15–93 years; mean 52 (sp 18) years) accepted to participate in the study; 267 were men (45·6%) and 319 women (54·4%). Participants' characteristics and stroke risk factors are shown in Table 9.1; 6·3% suffered from diabetes and 1·27% had a previous stroke. Almost 40% of the study population was overweight and almost 40% was normal weight.

(Continued)

Table 9.1 Characteristics and conventional stroke risk factors of the Aeolian population sample (*n* = 586), June 2007

Characteristic/risk factor	*n*	%
Age (years), mean	586	52
SD		18
Sex (male)	267	45.6
Migraine (and headache)	129	22.0
Hypertension	120	20.5
Smoke	84	14.3
Hypercholesterolaemia	79	13.5
Familiarity for stroke	67	11.5
Cardiopathy	19	3.2
Diabetes	37	6.3
Oral contraceptivies	19	3.2
Previous stroke	7	1.3
BMI		
Severe thinness ($<16.0 kg/m^2$)	1	0.2
Moderate thinness ($16.0–16.9 kg/m^2$)	3	0.4
Underweight ($17.0–18.5 kg/m^2$)	22	2.6
Normal weight ($18.5–24.9 kg/m^2$)	230	39.4
Overweight ($25.0–29.9 kg/m^2$)	220	37.4
Moderate obesity ($30.0–39.9 kg/m^2$)	108	18.5
Severe obesity ($\geq40.0 kg/m^2$)	7	1.3

Source: Eating habits in the population of the Aeolian Islands. An observational study, *Public Health Nutrition*, 22(9), 1590–6, reproduced with permission

1 Comment on the presentation of Table 9.1. What are the main features of the table?
2 How easy is it to cross-reference the data presented in the text with the table?
3 How could the table be improved?

· ·

· · · · · · **Develop your skills! 9.5** ·

The role of media within Western society, and indeed elsewhere, has become pervasive. Its reaches can be felt in all walks of life, and its impacts on human behaviour and how we live our lives are irrevocable. The abstract below is taken from a popular journal article, where the researchers undertook an observational study to investigate factors that influence media multitasking, when people are watching TV.

Look again at the article titled 'An observational study on how situational factors influence media multitasking with TV: The role of genres, dayparts, and social viewing' (at https://www.

Figure 9.5   Graphs and plots

Source: Photo by Myriam Jessier on Unsplash

tandfonline.com/doi/full/10.1080/15213269.2013.872038). Answer the questions related to this article, using the figure and accompanying text below to help you, if you cannot access the full journal article:

Differences across dayparts

Results also reveal a significant main effect of dayparts on media multitasking (see Table 9.2). The mean scores for each daypart are reported in Table 9.3 and displayed in Figure 9.6. The results suggest that the extent of media multitasking is the greatest in the morning (M D 1.21, SE D 0.006) and then in the afternoon (M D 1.19, SE D 0.005) with a significant difference between the two (p < 0.01). The extent of media multitasking in the evening (M D 1.14, SE D 0.005) is significantly lower (p < 0.01) than in the morning or afternoon. Therefore, Hypothesis 2 is supported.

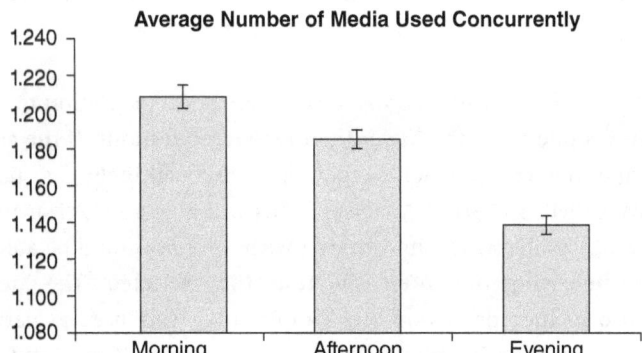

Figure 9.6   Average number of media used at different dayparts (with standard error bars)

Source: H.A.M. Voorveld and V. Viswanathan, An observational study on how situational factors influence media multitasking with TV: The role of genres, dayparts, and social viewing, *Media Psychology*, 18(4), 499–526 (2015), Taylor & Francis, reprinted by permission of the publisher

Note: The questions in this task do not require full access to the article or tables.

(Continued)

1　　Comment on the bar graph displayed in Figure 9.6. Is it clear? Are all the axes labelled?
2　　How well does the graph link to the accompanying text? Does the presentation flow well?
3　　Are there any improvements that could be made to the narrative in this excerpt from the paper?

. . . .

9.5 Evaluating the statistical language used in journal articles

The presentation of statistical analyses and data displays is ubiquitous, across a cacophony of journal articles. As we have discovered through this book, data need to be weaved into a coherent narrative that takes the reader on a journey. Using good statistical language is vital to achieve a compelling and clear argument, especially with reference to studies involving experiments or observational studies.

When trying to make sense of the statistical language presented, many of the skills and knowledge picked up in Chapter 4 will be vital to assist you in this task. Here are some pointers to help you:

1　　The abstract may give you some clues as to the type of study that was undertaken, for example if quantitative or qualitative approaches were used. This can help get your statistical thinking juices flowing to prepare you for the types of displays and methods you would expect to see.
2　　The introduction will give you the background behind why the research or study is important and may give some hints as to the population of interest. The research question may also be presented in this section of the journal article, which can give you a good indication as to the methods that are likely to be selected to answer them.
3　　The Methods section will give you a clear indication in relation to the approaches the authors undertook, to answer their research questions. If the authors decided to use experimental approaches, they may very well include a subsection under 'Methods' called 'Statistical Analyses'. This will give you a clear indication as to the types of variables they are working with, or you should be able to figure this out after reviewing the inferential tests they selected. The Methods section should also include details on any samples that were used as part of the study; they may also describe where the population come from, and, if applicable, further details on the sample (if they are people, or a particular cell line, of species of animal).
4　　The Results section is where you will see any statistical analyses being presented, as well as common data displays (bar graphs, dot plots, box plots, etc). This is likely to be the area where statistical language is presented (and will be picked up again most likely in the Discussion and Conclusion). Pay particular attention here to the use of the word 'significant' and look to see if the authors are referring to statistical

(which should include reference to *P*-values) or practical significance (which should be with reference to the confidence interval). You may also see the presentation of confidence intervals in a table, and then no mention in the explanatory text, so keep an eye out for this.

5   The interpretation of *P*-values can be presented in a multitude of ways, many of which are not correct. Look at Sections 4.4 and 4.5, in Chapter 4, for further guidance.

6   The Discussion (and possible Conclusion) section is where you are likely to see causal statements being made, attempting to explain a relation between variables. Make sure you fully scrutinise this section and reflect on the methods that were stated. For example, is a causal statement being made form an observation study? As we know from Chapter 8, this should be avoided.

Try to avoid:

1   Assuming that because a journal article has been published, the statistical analysis and language used will be correct. There are many examples of published work with statistics done badly!

2   Going too far and not believing anything! It can become easy, and dangerous, to become so sceptical that you do not trust or believe any results or conclusions from a published study.

· · · · · · Develop your skills! 9.6 ·

The journal article included in this task was published to help people navigate through other published work, mostly journal articles. This task will encourage you to evaluate the advice and suggestions presented.

Figure 9.7   Data presentation
Source: Photo by Mika Baumeister on Unsplash

(Continued)

Look at the article titled 'How to present research data?' https://www.ncbi.nlm.nih.gov/pmc/articles/PMC4453119/). Answer the questions related to this article, after reading this excerpt:

> Papers are often rejected because wrong statistical tests are used or interpreted incorrectly. A simple approach is to consult the statistician early. Bearing in mind that most readers are not statisticians, the reporting of any statistical tests should aim to be understandable by the average audience but sufficiently rigorous to withstand the critique of experts.

- Simple statistic such as mean and standard deviation, median, normality testing is better reported in text. For example, age of group A subjects was normally distributed with mean of 45.4 years old kg (SD = 5.6). More complicated statistical tests involving many variables are better illustrated in tables or graphs with their interpretation by text (see section on tables).

- We should quote and interpret p value correctly. It is preferable to quote the exact p value, since it is now easily obtained from standard statistical software. This is more so if the p value is statistically not significant, rather just quoting p>0.05 or p = ns. It is not necessary to report the exact p value that is smaller than 0.001 (quoting p<0.001 is sufficient); it is incorrect to report p = 0.0000 (as some software apt to report for very small p value).

- We should refrain from reporting such statement: 'mean systolic blood pressure for group A (135mmHg, SD = 12.5) was higher than group B (130mmHg, SD = 9.8) but did not reach statistical significance (t = 4.5, p = 0.56)'. When p did not show statistical significance (it might be >0.01 or >0.05, depending on which level you would take), it simply means no difference among groups.

- Confidence intervals. It is now preferable to report the 95% confidence intervals (95%CI) together with p value, especially if a hypothesis testing has been performed.

1   Do you think the four points presented are useful? Explain your answer.
2   Are there any parts of the excerpt that are confusing or not explained well? Why?
3   Which disciplines to you think the journal article is aimed at? Explain your answer.

···

····· Develop your skills! 9.7 ··· ·····································

The role of media within Western society, and indeed elsewhere, has become pervasive. Its reaches can be felt in all walks of life, and its impacts on human behaviour and how we live our lives are irrevocable. The abstract below is taken from a popular journal article we've already looked at, where the researchers undertook an observational study to investigate factors that influence media multitasking, when people are watching TV. Look again at the article titled 'An observational study on how situational factors influence media multitasking with TV: The role of genres, dayparts, and social viewing' (at https://www.tandfonline.com/doi/

Figure 9.8   Data displays

Source: Photo by Edward Howell on Unsplash

full/10.1080/15213269.2013.872038?casa_token=E60FzO0_dIlAAAAA%3ARjMAR6R7ZedxHpQFb-
BI8xbBD8-KYUm1VxBOJxlyexwviTzi547PqrqBfWT5Fp1pcNvN_G_9kXuX_sw). Answer the questions
related to this article. Use the figure and accompanying text below to help you if you cannot
access the full journal article (for these questions you don't need access to Table 9.2 or 9.3 that
are referenced in the excerpt, so don't worry if you can't see them).

Differences across dayparts

Results also reveal a significant main effect of dayparts on media multitasking (see Table
9.2). The mean scores for each daypart are reported in Table 9.3 and displayed in Figure
9.9. The results suggest that the extent of media multitasking is the greatest in the morning (M D 1.21, SE D 0.006) and then in the afternoon (M D 1.19, SE D 0.005) with a
significant difference between the two (p < 0.01). The extent of media multitasking in the
evening (M D 1.14, SE D 0.005) is significantly lower (p < 0.01) than in the morning or
afternoon. Therefore, Hypothesis 2 is supported.

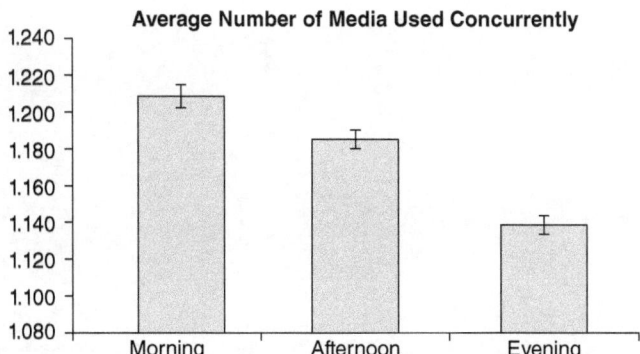

Figure 9.9   Average number of media used at different dayparts (with standard error bars)

Source: H.A.M. Voorveld and V. Viswanathan, An observational study on how situational factors
influence media multitasking with TV: The role of genres, dayparts, and social viewing, *Media
Psychology*, 18(4), 499–526 (2015), Taylor & Francis, reprinted by permission of the publisher

Note: The questions in this task do not require full access to the article or tables.

(Continued)

1 Comment on the statistical language used in the text. Is it used appropriately?
2 How could the language used in the text be improved?

....

Key points to remember

1 Not all journal articles will have the same subheadings as the ones presented in this chapter. For example, some may have subheadings like 'Background', instead of 'Introduction'.
2 Many journal articles get published with unclear and incorrect statistical analyses and language. There are many examples of published work with statistics done badly!
3 Try not to become too cynical to the point where you don't believe any results from published research!

....

References to support this chapter

Iyengar, S.S., and Lepper. M.R. (2000) When choice is demotivating: Can one desire too much of a good thing? *Journal of Personality and Social Psychology*, 79(6), 995–1006. https://doi. org/10.1037//0022-3514.79.6.995

La Spina, P., Savica, R., Ciacciarelli, A., Cotroneo, M., Dell'Aera, C., Grillo, F. ... Musolino, R. F. (2018) Eating habits in the population of the Aeolian Islands: An observational study. *Public Health Nutrition*, 22 (9), 1590–1596. https://doi.org/10.1017/S1368980018003397

Tong, F.S. and Aziz, A.F.A. (2006) How to present research data? *Malaysian Family Physician: The Official Journal of the Academy of Family Physicians of Malaysia*, 1 (2–3): 82–85. http://www.ncbi. nlm.nih.gov/pmc/articles/PMC4453119/

Voorveld, H.A.M. and Viswanathan, V. (2014) An observational study on how situational factors influence media multitasking with TV: The role of genres, dayparts, and social viewing. *Media Psychology*, 18 (4), 499–526. https://doi.org/10.1080/15213269.2013.872038

....

10

TELLING STORIES WITH DESCRIPTIVE DATA, TABLES, AND GRAPHS

An overview of this chapter

The communication of information is ubiquitous; however, knowing the most effective way to present and explain data can be a challenge. Weaving data into a story well can leave a strong impression on the reader, convincing them of a particular argument or stance. Data stories are often overlooked by authors, including the careful selection and presentation of data into a coherent narrative. This chapter will provide a rationale as to why data stories are useful, leading to some guidance for presenting data stories, including the importance of knowing the intended audience. The advantages of using descriptive data are also discussed, along with information on using graphs and tables.

Resources to support this chapter

This chapter draws on a range of sources to help show you what makes a good data story. In this chapter, we use real-world examples to show you what good and bad data stories look like. You can build your skills by trying the suggested activities on presenting data clearly and in a coherent fashion.

Resource	Date accessed	Location
Big data in media and telco: 6 applications and use cases, *Talend*, 2021	17 August 2021	http://web.archive.org/web/20230126214215/https:/www.talend.com/resources/big-data-media-telco/
SENSITIVE DATA: Covid-19 news and information: consumption and attitudes – interactive data, *Ofcom*, 2021	17 August 2021	https://www.ofcom.org.uk/research-and-data/tv-radio-and-on-demand/news-media/coronavirus-news-consumption-attitudes-behaviour/interactive-data

(Continued)

(Continued)

Resource	Date accessed	Location
SENSITIVE DATA: Fake news: What's real? What's distortion? *BBC*, 2021	17 August 2021	https://www.bbc.co.uk/news/topics/cjxv13v27dyt/fake-news
Almost 50 shops a day disappear from High Streets, *BBC*, 2021	20 October 2021	https://www.bbc.co.uk/news/business-58433461

The websites referred to in all the activities are provided as part of this book's online resources. You can find them at **https://study.sagepub.com/jonesstatsliteracy**.

10.1 The value of a good data story and the Pixar method

The need for good data stories is only going to increase in the future, as we collect and store more and more data on a variety of topics. With the shift towards more self-service capabilities in analytics and business intelligence, the pool of people generating insights will expand beyond just analysts and data scientists. It will become everybody's responsibility to be versed in not only navigating, but also creating engaging succinct and persuasive data stories.

Making a data story compelling and engaging can include a variety of techniques and methods. Stories in general have been told for centuries, and there are frameworks and methods to help create a compelling narrative. The Pixar method for telling stories is a proven approach that has captivated audiences worldwide. At the heart of this method is the understanding that storytelling is a powerful tool for connecting with emotions and creating compelling narratives. These techniques can also be used when telling stories with data and assist in creating a coherent data narrative. Pixar emphasises the importance of crafting relatable characters, which can also be extended to include interesting topics and contexts, with depth and complexity, allowing readers to form an emotional bond with these areas. The method also emphasises the significance of a strong story structure, employing the classic three-act structure to establish a clear beginning, middle and end. Additionally, Pixar encourages a collaborative and iterative process, where stories are refined through constant feedback and revision. This method values the exploration of various ideas and perspectives, pushing storytellers to take risks and challenge conventions. Ultimately, the Pixar method highlights the essence of storytelling: the ability to transport readers to imaginative worlds, evoke a range of emotions, instil curiosity and leave a lasting impact.

Numbers also have an important story to tell and can be used with the Pixar method. The way numbers are communicated, presented and folded into a coherent narrative requires a multitude of skills in order to be done well. Some of these skills you have already picked up throughout this book. These skills rely on you to give the numbers a clear and convincing voice. Any insight worth sharing is probably best shared as a data story. The phrase **data storytelling**, or telling stories with data, has been associated with many things, among them data visualisations, infographics, dashboards, and data presentations. Too often, data storytelling is interpreted as just visualising data effectively, but it is much more than just creating visually appealing data charts. Data storytelling is often a structured approach to communicating data insights, and it involves a combination of three key elements: data, visuals, and narrative.

········· Develop your skills! 10.1 ···· ···

Big data is everywhere! It is becoming increasingly available from governments, businesses, and social media. Big data has had a profound impact on the way we live our lives and will continue to do so as technology and the way we collect data improve.

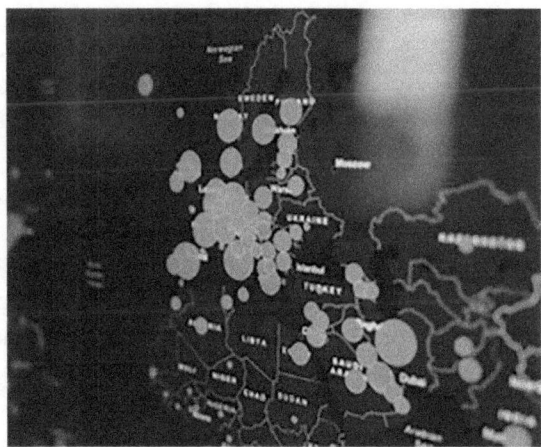

Figure 10.1   Big data
Source: Photo by Clay Banks on Unsplash

Look at the webpage titled 'Big data in media and telco: 6 applications and use cases' (at http://web.archive.org/web/20230126214215/https:/www.talend.com/resources/big-data-media-telco/) and answer the following questions:

1   Firstly, what does 'big data' mean to you?
2   Summarise the advantages of big data in the media, entertainment and telecommunications.

··

10.2 Sensitive data

As we saw in the introduction to this chapter, we live in a world full of exciting and engaging subjects that are all underpinned by data. Data can come in many different forms and can be used in a variety of contexts. The same data can also be used to explore many different contexts. Context and subject-matter can give data a new lease of life, used to convey a message or set of ideas. Different contexts and subjects can mean different things to different people, both on an individual level and to different cross-sections of society. Naturally, the use of sensitive data sets can be linked to controversial or sensitive topics, which certain groups of people may want to avoid. We therefore need a definition of 'sensitive data'.

> **Sensitive data** is defined as any data set that is assigned to a controversial or potentially sensitive topic, that can be emotionally triggering, upsetting·or cause distress.

To define 'sensitive data' is not easy, since it can encompass so many kinds of data, across a multitude of contexts and topics. It therefore feels appropriate to assign a broad definition to sensitive data.

So many controversial or sensitive topics will often involve cross-sections of society, which can come from different cultures. They can describe or explain the way we live, how people identify themselves, and they can also touch on social norms and beliefs. They can define the way we live and interact with others. It is therefore *essential* that we collect and use data on these topics to better understand ourselves, and society. To censor and exclude such topics distorts the truth and can give a false image of the world we live in.

Learning that data are important to help us explain the world around us can support the development of a more rounded understanding of who we are and, our own position in it, and can help us to challenge our own beliefs and ideologies. It can also foster a deeper understanding of other people's perspectives, make us more empathetic, and lead us to delve deeper into epistemological and ontological considerations.

However, sensitive data can be emotionally triggering, upsetting, and too difficult for some people to deal with.

10.2.1 Framework for dealing with sensitive data

The following points can be used to help you to navigate sensitive data, as well as helping to develop our critical thinking skills, ensuring you can maximise your learning opportunities:

1   What are the 'sensitive data' concerned with? For example, are they to do with gender diversity? Or religion? Ask yourself: how does this topic make you feel? Does it mean something personal to you? At this point, if you feel it is emotionally triggering or too sensitive a topic for you to manage, it might be better for you to skip it.

2   If you feel comfortable with the topic, next ask yourself whether the article is interesting. Try to think about what it can tell us about the world we live in. What issue is it tackling or getting you to think about? Your tutors and lecturers might already have shown you bad examples of reports or pieces of work relevant to what you are learning. And there is no shortage of these in the world! Engaging with

exercises like this helps you to develop your critical thinking skills and can also nurture your abilities to distinguish between what is deemed to be good, and what is not so good.

3   When looking at the data, think about why they were collected, and how they are being used in the report or story. What is the agenda of the author of the report or story you are looking at, and can this be used to explain their perspectives or views being portrayed?

4   Finally, think about the intended audience of the report or story. This can really help you to understand the language a piece of work uses to deliver its key messages. It can also help you to unearth the reasons why the authors have taken a certain angle, and whether you think they have a case, that is, how persuasive is the piece of work being presented.

········ Develop your skills! 10.2 ·· ··

SENSITIVE DATA EXAMPLE: Covid-19

Covid-19 has had a huge global impact on the way we live our lives. This task will explore several reports about Covid-19.

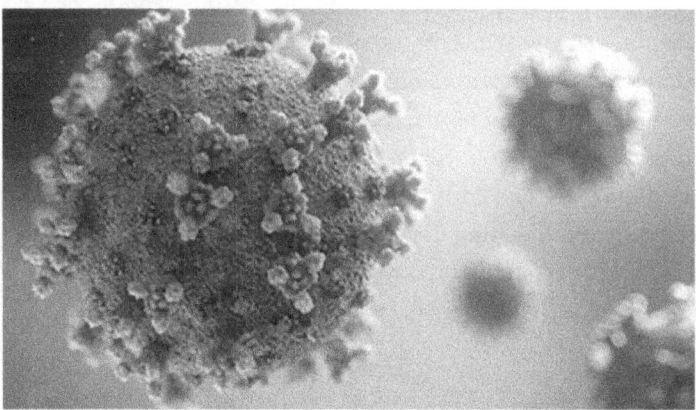

Figure 10.2   Visualisation of Covid-19
Source: Photo by Fusion Medical Animation on Unsplash

Look at the Ofcom webpage titled 'Covid-19 news and information: consumption and attitudes – interactive data' (at https://www.ofcom.org.uk/research-and-data/tv-radio-and-on-demand/news-media/coronavirus-news-consumption-attitudes-behaviour/interactive-data), which you first encountered in Chapter 7, and answer the following questions:

(Continued)

1   What are your impressions of this webpage?
2   Is there anything that surprises you on the landing page (i.e., the initial page you land on
     when you click on the link related to this task)?
3   Explore several webpages linked to this data platform. What types of data do you see? How
     are they presented? What types of figures are there?
4   Reflecting on your answers to Question 3, do you think most people would be able to make
     sense of the data you have looked at, and the way they are presented? Explain your answer.

. . . .

10.3 Breathing life into data stories

Breathing life into stories is the art of infusing them with vitality, depth and a sense of
authenticity. It involves going beyond the mere recitation of events and delving into
the essence of characters, settings, contexts and themes. It requires crafting rich and
vivid settings that transport audiences to immersive worlds. Additionally, breathing life
into stories entails exploring complex themes and weaving them into the narrative
fabric, addressing universal human experiences and challenging perspectives. It can
involve employing vivid and evocative language, engaging all the senses to paint a vivid
picture in the minds of the audience. Ultimately, breathing life into stories is about tap-
ping into the power of storytelling to captivate, inspire, and leave a lasting impact,
ensuring that the tales we tell become living entities that have the potential to connect
deeply with people's hearts and minds.

Creating a data story requires the use of multiple skills to ensure there are harmoni-
ous connections between your main body of text and the graphs and tables used to
draw the reader to specific points of interest, central to the big ideas in the data story.
Here are some tips for bringing data stories to life. The language you use will depend
greatly on the intended audience. For example, if it's an academic piece of work, more
advanced or technical language may be required.

Use:

1   Engaging language that people understand
2   Short sentences, short paragraphs
3   Simple language: 'get', not 'acquire'; 'about', not 'approximately'; 'same', not
     'identical'
4   Bullet lists for easy scanning
5   Numbers in a consistent fashion: for example, choose 40 or forty, and stick with the
     choice made
6   Rounded numbers (by rounding both long decimals and big numbers)

Try to avoid:

1 'Elevator statistics': this went up, that went down
2 Jargon and technical terms
3 Using large numbers, which can be difficult to grasp. Use the words 'million', 'billion' and 'trillion'. Instead of '16,956,990' write 'about 17 million'.
4 Using percentages for small numbers (i.e., sample or population sizes that are less than 50)
5 All capital letters and all italics: mixed upper and lower case is easier to read

········Develop your skills! 10.3 ···

SENSITIVE DATA EXAMPLE: Fake news

Fake news has become increasingly common in society, especially on social media platforms. How do we determine what is real and what is fake? Many of the skills you have developed in using this book will help you to do this; however, this task looks specifically at fake news.

Figure 10.3 Fake news
Source: Photo by Mika Baumeister on Unsplash

(Continued)

Look again at the BBC 'Fake news' webpage (at https://www.bbc.co.uk/news/topics/cjxv13v27dyt), which you first met in Chapter 7, then answer the following questions:

1 How do you spot fake news? What makes you question the accuracy and factual basis of an account or statement, for example in a social media post?
2 Why do you think fake news exists? Use a specific example (using the link for this task) to answer this question.
3 What makes fake news items convincing?

· ·

10.4 Using tables and graphs

Data are often presented in tables to highlight interesting patterns or relationships, as we have explored in Chapter 4 and elsewhere in this book. Being able to read tables, spot patterns of interest, and then communicate those findings are extremely valuable skills, applicable across a range of disciplines (Gal, 2002). Tables can include data involving percentages, proportions or ratios, or the actual values from the variables themselves (i.e., distance between participants' homes and schools). If percentages are used, it's a good idea to include the sample or population sizes in the table. Also, if numeric data are to be included, think about the level of accuracy that should be used. For example, reporting the number of children adopted in a certain region in 1932 could include the use of whole numbers, or mean values that could include decimal places.

Good tables should complement and be connected to text. They should present numbers in a concise and organised fashion to support the main points being made. Tables help minimise numbers in the data story (i.e., the main body of text). Tables shouldn't be too large or complex. One decimal place will be adequate for most data. In specific cases, however, two or more decimal places may be required to draw attention to subtle differences in the data. Tables should also contain a title that is clear and concise.

You have come across tables back in Chapter 3. As a reminder, Table 10.1 is a two-way table of counts of crimes committed in San Francisco. As a table, it is clear, concise, with appropriate columns and row headings, and includes a title.

Table 10.1 Two-way table of counts: crime levels in San Francisco, 2014

	Theft	Assault	Total
AM	23,275	6,521	29,796
PM	56,082	14,122	70,204
Total	79,357	20,643	100,000

Displaying data in graphical format is a common way of presenting data, found in many disciplines. Whenever presenting a graph, the following questions can help to ascertain the point of them, and the intended messages to be conveyed:

1   What are the main features of the graph?
2   What other details are useful for understanding the variable?

The type of graph used will depend on the number and type of variables being included. Table 10.2 can assist with making these decisions (this is a recap from Chapter 4). An effective graph has a clear, visual message, with an informative and concise heading.

Table 10.2   Strengths and weaknesses of commonly used graphs

Type	Strengths and weaknesses	Example
Dot plot	Retains numerical information, can also check for skewed data (data that can be bunched to the left or right of the graph)	Gender (categorical) and salary (numeric)
Scatter plot	• Can give you useful information on whether the axis in the graph is associated with each other • Retains numerical information	Age and salary (numeric)
Box-and-whisker plot	• Very good for comparing numeric data subdivided by categorical data • Displays centre and spread • Not useful for small data sets	Gender (categorical) and salary (numeric)
Histogram	• Displays relative density of observations (i.e., gives a good idea of the shape of the distribution) • Good for large amounts of data	Heights (numeric)
Bar chart	• Good for displaying categorical data, either as proportions for a single variable, or more • Can quickly reveal interesting patterns • Often don't know the sample size (unless using specific types of software)	Marital status (categorical)

Good statistical graphics (or plots) should:

1   Present logical visual patterns that are not too busy, or contain too much information
2   Have the appropriate scale on the axis. For example, percentage increases over time can be big or small values. Making the scale very small (i.e., percentage increments of 0.1%) would make the percentage increase look a lot bigger than it is

Achieve clarity in graphs (or plots) by:

1 Using data values on a graph only if they don't interfere with the reader's ability to see patterns of interest in the data (i.e., the message that you really want to convey)
2 Making all text on the graph easy to understand
3 Not using abbreviations
4 Avoiding acronyms
5 Avoiding legends except on maps

You have come across several examples of clear and well-presented graphs in this book already. Figure 10.4 (presented in Chapter 3) is a side-by-side dot plot, with box-and-whisker plots underneath each. The data has been subdivided by gender. The figure has a clear title, labelled axis, and has the mean included, represented by a triangle. The scale on the *x*-axis is sensible, and wide enough to see the data point spread, to enable to reader to look for interesting patterns (i.e., the data points are not bunched up, or overprinted (on top of each other)).

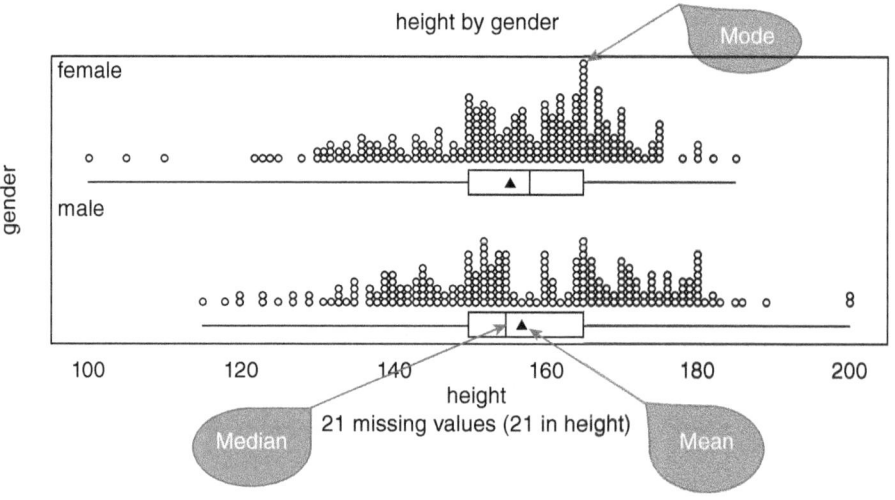

Figure 10.4 Central tendency and measures of centre and spread

···· ··· Develop your skills! 10.4 ·· ··· ··································

This task provides a fictional data story released by a newspaper (based on real events) that has several problems with it. These are listed underneath this media report.

DOES ANYONE CARE ABOUT BREXIT ANYMOR? THE PERILS OF TOO MUCH MEDIA ATTENTION AND THE IMPACT THIS IS HAVING ON IMPORTANT ISSUES

The people of Great Britain are proud and resourceful, and overall, very tolerant. The demographics of the country vary greatly, across the four countries of this kingdom. Protests have erupted since the Brexit vote came to pass bac in June 2016.

Public opinion on the vote for the UK to leave the EU has gone up and down since the official vote back in June 2016. It appears that there is a greater proportion of remain voters in a recent poll taken in **July 2019,** with the leave amp now in the minority.

Many people feel like they were lied to by the leave campaign before the official vote took place. But the answer now remains do people still care about this important issue? Has it become too political?

A recent poll conducted by WUUKL asked a sample of people in London, 'do you still care about Brexit?' revealed that 71.5578% said **No**, with the remainder saying **Yes or Don't** know.

So why do people in the UK no longer care about this important political and societal issue?

There are a number of issues with this article:

1 The heading is too long and in capital letters
2 Some words are underlined for no clear reason
3 Paragraphs are not arranged clearly
4 Exact dates of polls and when the official vote was reported are not mentioned
5 Some words are bolded for unknown reasons
6 Acronyms are used, with some less familiar than others
7 Elevator statistics have been used: UK to leave the EU has gone up and down since the official vote back. It is not clear what the actual figures are
8 The first paragraph is unclear and clumsy
9 The text needs to be proofread – multiple spelling mistakes are present
10 Too many decimal places have been used in the percentage reported
11 It is unclear what the split is between No and Don't know
12 Too many questions are posed in general, not enough answers

Try improving this article yourself, using the above points to guide you.

· · · · · · Develop your skills! 10.5 · · ·

Shopping forms an integral part of many of our lives. It is linked to social, cultural and religious events, steeped in a history full of trade and exchange for goods and money. This task will encourage you to think about the current economic climate in Western society and explore the continued closure of shops on the high street in the UK.

(Continued)

Figure 10.5 Local café store closure
Source: Photo by Tim Mossholder on Unsplash

Read the article titled 'Almost 50 shops a day disappear from High Streets', 2021 (at https://www.bbc.co.uk/news/business-58433461), which you first encountered in Chapter 7, then answer the following questions:

1   Does the article present a balanced view? Is it biased or leaning towards one perspective or point of view? Explain your answer.
2   How does the article use graphs? Are they easy to understand? Are they presented well? Provide examples from the article to support your answer.

∙∙

Key points to remember

1   Think clearly in terms of what makes a good data story. Is it engaging? Does it contain facts that are accurate and presented in an understandable way? Is there a clear and coherent narrative?
2   Think about who the audience potentially is when writing a data story. This can help guide and frame the work you are undertaking.
3   Pay close attention to several of the checkpoints in the chapter and develop your writing to incorporate the good practices and guidance presented.
4   Use the guidelines in the sensitive data section (see Chapter 1) and elements of Chapter 8, as a checklist to help you to think critically about data stories, their purpose, and their accuracy, validity and reliability.

Further reading to support this chapter

Bleakley, A. (2005) Stories as data, data as stories: Making sense of narrative inquiry in clinical education. *Medical Education*, 39(5), 534–40. https://onlinelibrary.wiley.com/doi/full/10.1111/j.1365-2929.2005.02126.x

Few, S. (2004) *Show Me the Numbers: Designing Tables and Graphs to Enlighten*. Oakland, CA: Analytics Press.

Kosslyn, S.M. (1994) *Elements of Graph Design*. San Francisco: W.H. Freeman and Co.

Martin, N. (2018, November) Data visualization: How to tell a story with data. *Forbes*. https://www.forbes.com/sites/nicolemartin1/2018/11/01/data-visualization-how-to-tell-a-story-with-data/#6103e81a4368.

Miller, J.E. (2004) *The Chicago Guide to Writing about Numbers*. Chicago: University of Chicago Press.

Ploszajski, A. (2012) *Hand Made: A Scientist's Search for Meaning through Making*. London: Bloomsbury Sigma.

Sebel, E. and Heer, P. (2011) Narrative visualization: Telling stories with data. *IEEE Transactions on Visualization and Computer Graphics*, 16(6), 1139–48. http://vis.stanford.edu/files/2010-Narrative-InfoVis.pdf.

Yorke, J. (2014) *Into The Woods: How Stories Work and Why We Tell Them*. London: Penguin.

References to support this chapter

BBC (2021) Fake news: What's real? What's distortion?. https://www.bbc.co.uk/news/topics/cjxv13v27dyt (accessed 17 August 2021).

Gal, I. (2002) Adults' statistical literacy: Meanings, components, responsibilities. *International Statistical Reviews*, 70(1), 1–51.

NatCen Social Research (2019, 18 November) What UK thinks EU: Non-partisan information on UK attitudes to the EU before and since the EU Referendum. https://web.archive.org/web/20191216023219/https://whatukthinks.org/eu/questions/if-there-was-a-referendum-on-britains-membership-of-the-eu-how-would-you-vote-2/

Ofcom (2021) Covid-19 news and information: Consumption and attitudes – interactive data. https://www.ofcom.org.uk/research-and-data/tv-radio-and-on-demand/news-media/coronavirus-news-consumption-attitudes-behaviour/interactive-data

Simpson, E. (2021) Almost 50 shops a day disappear from High Streets, *BBC*. www.bbc.co.uk/news/business-58433461 (accessed 20 October 2021).

Talend (2021) Big data in media and telco: 6 applications and use cases. http://web.archive.org/web/20230126214215/https:/www.talend.com/resources/big-data-media-telco/ (accessed 17 August 2021).

11

COMMON MISCONCEPTIONS IN STATISTICS AND STATISTICAL LITERACY

An overview of this chapter

This chapter links directly to the previous chapters in the book (especially the previous two), presenting you with several common misconceptions in statistics. These misconceptions will be highlighted using a range of examples, including published journal articles, social media platforms and official reports.

Resources to support this chapter

This chapter draws on a range of sources to help show you the common misconceptions in statistics. In this chapter, we use a range of real-world examples to show you examples of statistics communicated badly. You can build your skills by trying the suggested activities in which you can scrutinise descriptive data displays, evaluate the ways in which data are collected, and in relation to discussing and communicating data.

Resource	Date accessed	Location
How to present research data?, *Malaysian Family Physician*, 2006	11 August 2022	https://www.ncbi.nlm.nih.gov/pmc/articles/PMC4453119/
SENSITIVE DATA: Fake news, *BBC*, 2022	11 August 2022	https://www.bbc.co.uk/news/topics/cjxv13v27dyt/fake-news
There's science behind your inexplicably close relationship with your cat, *Quartz*, 2014	09 June 2022	https://qz.com/304236/theres-science-behind-your-inexplicably-close-relationship-with-your-cat/

(Continued)

(Continued)

Resource	Date accessed	Location
Open letter to the Kansas School Board (2006), *Church of the Flying Spaghetti Monster*, 2022	11 August 2022	https://www.spaghettimonster.org/about/open-letter/
Statistical tests, P values, confidence intervals, and power: A guide to misinterpretations, *European Journal of Epidemiology*, 2016	11 August 2022	https://www.ncbi.nlm.nih.gov/pmc/articles/PMC4877414/

The websites referred to in all the activities are provided as part of this book's online resources. You can find them at **https://study.sagepub.com/jonesstatsliteracy**.

11.1 Misconceptions in relation to collecting data

Statistical literacy uses a range of thinking and practical skills that include knowledge, comprehension, application, analysis, synthesis and evaluation. It enables a feel for data, including being able to support an argument with evidence, but also being aware of the variety of interpretations that are possible from those data. A statistically literate person will appreciate that information and conclusions drawn from data will have uncertainty attached to them.

One misconception worthy of note involves the following erroneous statement: it does not matter how the data were collected. Throughout this book, we have looked at multiple methods for collecting data (see Chapter 8). We have also explored important perspectives, linked to ontological and epistemological considerations, as well as factors that can impact the quality of data, their reliability, validity, and the type of inference that can be made (Chapters 3 and 8–10). These key points should emphasise the importance for us of reflecting on how data were collected, since it has so many other knock-on effects and impacts.

··· ··Develop your skills! 11.1 ··· ······

SENSITIVE DATA EXAMPLE: Religion and faith

This task will enable you to build your skills in being able to critically evaluate an interesting and engaging data story, based on religion.

Figure 11.1   Monster in the forest

Source: Photo by Mariusz Słoński on Unsplash

Look at the webpage titled 'Open letter to the Kansas School Board (2006)' (at https://www. spaghettimonster.org/about/open-letter/), then answer the following questions:

1   What is the article about?
2   Do you think the article is being serious? Give reasons for your answer.
3   What point do you think the article is trying to make?
4   Comment on the evidence and data presented in the article. How is it used to present an argument, or certain point of view?

· ·

· · · · · · · Develop your skills! 11.2 ·

SENSITIVE DATA EXAMPLE: Fake news

This task will enable you to further build your skills in evaluating an interesting and engaging data story, on fake news.

(Continued)

Figure 11.2   Fake news
Source: Photo by Kajetan Sumila on Unsplash

Look at the BBC 'Fake news' webpage (at https://www.bbc.co.uk/news/topics/cjxv13v27dyt), which you first saw in Chapter 7, then answer the following questions:

1   Click on one of the stories on the webpage and read the article.
2   Make notes on the news story. Is it eye-catching? Does it take long to read? How are the data displayed in news story?
3   What is it that the news story is tackling, related to fake news?
4   Are there any features from the news story you have selected that make you trust what you have read? Describe them.
5   Are there any clues in the news story that give an indication as to how the supporting data were collected or obtained? Explain your answer.

· ·

· · · · · · Develop your skills! 11.3 ·

Figure 11.3   A cat relaxing
Source: Photo by Zeke Tucker on Unsplash

Look at the webpage titled 'There's science behind your inexplicably close relationship with your cat', 2014 (at https://qz.com/304236/theres-science-behind-your-inexplicably-close-relationship-with-your-cat/), which you first saw in Chapter 7, then answer the following questions:

1   Comment on the different forms of evidence and data presented in the article. Include an analysis of the validity and reliability of the data presented.
2   Are the arguments presented compelling? Use your comments in response to Question 1 to support your answer.
3   Comment on the language used in the article, especially with reference to the data presented. Would you say the author is statistically literate? Explain your answer.

· ·

Try to avoid:

1   Overlooking the importance of how data were collected.
2   Believing claims at face value. Keep asking questions about the legitimacy of the data being presented, along with the arguments being made.

11.2 Misconceptions in relation to descriptive data

Averages are often used in social media reports, news stories and journal articles. They are one of the most widely used methods to report data, reducing the data points to a single value. The standard deviation gives us more of an idea of the variability present in the data. This is not used as much in social media reports and in news stories. This section discusses misconceptions concerning the use of averages and the standard deviation.

11.2.1 Averages

When descriptive data are reported in the media, journal articles or other outlets, the average value is often included. Most people use the word 'average', in relation to numbers, without thinking about how the numbers are distributed. People often do not understand that averages can be extremely misleading. People think that 'average' is synonymous with 'normal' or 'most common'. However, as we discovered in Chapter 3, the mean is often what people are talking about when they refer to average values. However, the word 'average' does not always refer to the mean. A person looking at statistical data needs to know what the underlying data look like to really understand them, which can be facilitated by using a graph like a dot plot. A skewed distribution, which can sometimes result from just one number being much larger or much smaller than the rest, may well be poorly represented by the mean. This is because the mean takes into consideration every single number in the data set, so the extreme values will push the mean away from the centre of the data set. In cases like this, the median should be used, as this takes into consideration only the positions of the values in the ordered data set.

11.2.2 Standard deviation

The standard deviation is also overused at the expense of other measures of dispersion. For example, its use with the arithmetic mean (as mean ± SD) is misleading for data with a skewed distribution. Similarly, neither the mean nor the standard deviation is appropriate for summarising ordinal variables. For skewed or data from ordinal variables, a box-and-whisker plot containing a five-number summary (minimum, lower quartile, median, upper quartile and maximum) is much more informative. Researchers often admit that the distribution of their data is not remotely normal, and consequently use a nonparametric statistical test to compare medians, yet still (misleadingly) present their results as means with standard deviations.

········· Develop your skills! 11.4 ···

This task will enable you to further build your skills in being able to evaluate a data story, using an article that tries to explain how to present research data.

Figure 11.4   Sugar and diabetes
Source: Photo by Myriam Zilles on Unsplash

Look at the article titled 'How to present research data?', 2006 (at https://www.ncbi.nlm.nih.gov/pmc/articles/PMC4453119/), which you first met in Chapter 9, then answer the following questions:

1   Read the first few lines of the 'Text' section. These give two ways of presenting a mean value and explain why the first is not as good as the second. Do you agree with the authors' explanation? Give reasons for your answer, referring to the language used.
2   Does anything surprise you about the article? What is missing from the beginning part, that is usually included in all journal articles?
3   Read the 'Some General Rules' section. Comment on the language used. Is it clear? Could it be improved? Explain your answer.

··

Try to avoid:

1   Assuming that people, in general, understand statistical terms like average and standard deviation.
2   Underestimating the value of reporting averages and standard deviations.
3   Using the mean and standard deviation inappropriately (e.g. for categorical variables).

11.3 Misconceptions in relation to discussing and communicating data

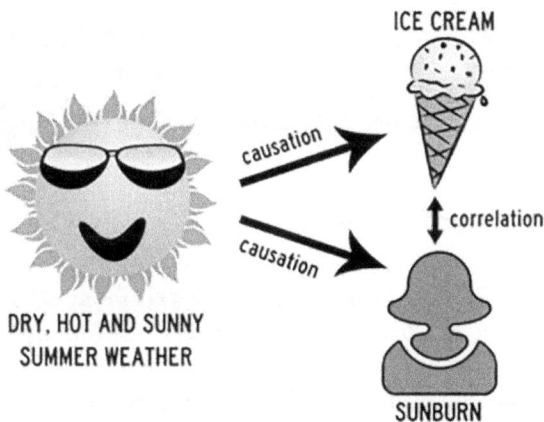

Figure 11.5   Spurious correlation example

11.3.1 Causation and correlation

When looking at correlated data (Chapter 4), we can often say that there is an association between variables, for example as one variable increases so does another, and if the data points related to these variables can be modelled using a straight line, we may call it a linear relationship. This relationship might lead us to assume that a change to one thing *causes* a change in the other. Reasons for this include a human need to simplify information, so we can make sense of it. Our brains often do that by making assumptions about things based on slight relationships or bias. But that thinking process is not foolproof. An example is when we mistake **correlation** for **causation**. For example, in Figure 11.5, sunburn cases and ice cream sales are correlated. Bias can make us conclude that one thing must cause another if both change in the same way at the same time. There are many examples of interesting correlations that are unlikely to have any causal links, which are called spurious correlations. The following task covers this topic in more detail.

········ Develop your skills! 11.5 ·· ···

This task will explore several examples of spurious correlations, which will help to build your critical thinking skills and confidence with statistical literacy.

Figure 11.6   Lego pirates
Source: Photo by Tom Briskey on Unsplash

Take another look at the webpage titled 'Open Letter to the Kansas School Board (2006)' (at https://www.spaghettimonster.org/about/open-letter/), then answer the following questions:

1   Look at the plot labelled 'Global Average Temperature Vs. Number of Pirates'. Comment on the presentation of the plot. Are there any issues with it?
2   How do you think the data were collected for the variables listed on this plot?
3   Describe the relationship between the variables in this plot.
4   Do you think there is a causal link between the variables in the plot?
5   The author makes the following claim: 'As you can see, there is a statistically significant inverse relationship between pirates and global temperature.' Do you think this claim is justified? Explain your answer.

···

11.3.2 Regression to the mean

Regression to the mean (RTM) is a statistical phenomenon that can make natural variation in repeated data (i.e. obtained by taking multiple measurements of the response variable, for example investigating the effect of chocolate on heart rate, where multiple measurements of heart rate are taken) look like real change. It happens when unusually large or small measurements, which can include baseline observations, are followed by measurements that are closer to the mean (Figure 11.6). To help identify RTM, subjects or participants in an experiment should be randomly allocated to treatment groups (e.g., groups receiving chocolate versus groups receiving no choc-olate, with the response variable being heart rate) the responses from all groups should be equally affected if the phenomenon (i.e., RTM) exists in the data. Remember we came across random allocation in Chapter 8, under experiments.

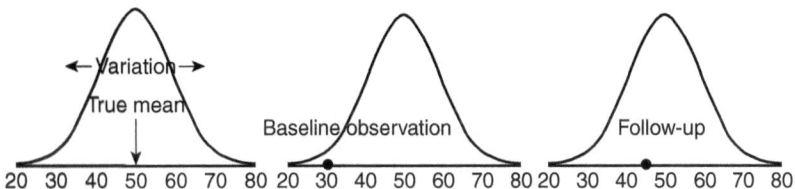

Figure 11.7   Regression to the mean example

11.3.3 Significance and *P*-values

There are many inaccurate ways to report and interpret *P*-values, and misconceptions abound. In Chapter 5 we looked at what a *P*-value represents, and some of the issues that can arise when these values are reported. We also looked at the difference between statistical significance (which can be ascertained using the *P*-value) and practical sig-nificance (which can be ascertained from looking at the confidence interval). Remember that the *P*-value represents the strength of evidence we have against the null hypothesis; this should be communicated alongside the confidence interval to give an idea of the size of any potential observed differences. The language used to communicate these complex ideas is so important and often incorrect, and there are plenty of examples in journal articles across the sciences and other disciplinary areas.

Sometimes the *P*-value is presented as the probability of the null hypothesis being correct, or it is somehow a measure of the probability of independence or no association between variables. These interpretations are incorrect. Another incorrect assumption made by many scientists and researchers is that a large *P*-value indicates that there is no treatment effect (in the context of an experiment) or no difference between means. Remember, *P*-values only give us evidence against the null hypothesis and enable us to rule out chance acting alone (if the *P*-value is small, usually smaller than 0.05). What we cannot do with a *P*-value is indicate whether there is or is not a treatment effect, in the context of an experiment.

It's also important to remember that the *P*-value should not be viewed as a set or fixed value that cannot be changed. Sadly, many disciplines see the *P*-value as being either statistically significant or not, so a binary outcome (for example, many use 95%, looking for *P*-values smaller than 0.05). As we have seen in Chapter 5, it is more com-plicated than this.

········ Develop your skills! 11.6 ··

This task will enable you to further develop your skills in evaluating *P*-values, as well as reviewing their interpretation and misinterpretation.

Figure 11.8   Confused

Source: Photo by Jeshoots.com on Unsplash

Look at the article titled 'Statistical tests, *P* values, confidence intervals, and power: A guide to misinterpretations', 2016 (at https://www.ncbi.nlm.nih.gov/pmc/articles/PMC4877414/), then answer the following questions:

1   Read the abstract. Why do you think *P*-values are not interpreted and communicated correctly in journal articles and other outlets?
2   How can the issue highlighted in Question 1 be tackled, more broadly? Explain your answer.
3   Look at first paragraph of the Conclusion section. What is the main caution presented here?

···

11.3.4 Inference and sample size

Common misconceptions concerning inference and sample size often include sampling and non-sampling errors as well as inference errors (covered in Chapters 4 and 6). Data collected using a sample, especially in a poll or survey, are often used to make claims about the population they came from. TV adverts often use very small sample sizes to communicate views or levels of agreements about the effectiveness of a product, to persuade viewers to purchase it. The sample chosen is small and may not be representative of the population it came from, leading to sampling errors. In addition, if the sample is taken from a specific area, and claims from the data are extended to other areas, then this would represent a non-sampling error, specifically the transferring of findings.

Try to avoid:

1   Assuming correlation is the same as causation. Remember they are different phenomena, and the existence of a correlation does not mean there is an association between the variables involved.
2   Confusing statistical and practical significance.
3   Misusing findings from a statistical test, and subsequently misinterpreting or overstating a P-value.
4   Making sampling or non-sampling errors and carrying forward these errors when communicating findings from data.

Key points to remember

1   Details of how data were collected can provide important information that can help when interpreting data and assessing their validity.
2   Do question claims that are made in journal articles social media and news platforms. Keep asking questions about the legitimacy of the data being presented, along with the arguments being conveyed.
3   Don't assume that people, in general, understand statistical terms like average and standard deviation.
4   Reporting averages and standard deviation values can help present a compelling argument, when reporting findings from data.
5   The mean and standard deviation should only be used with variables of the appropriate type, and not, for example, with categorical variables. Sometimes the median and interquartile range will be a better summary of data than the mean and standard deviation, for example when the data is not normally distributed (bell-shaped).
6   Correlation is not same as causation. Remember they are different phenomena, and the existence of a correlation does not mean there is an association between the variables involved.
7   Keep in mind the differences between statistical and practical significance, and the value of each.
8   Keep an eye out for the misuse of findings from a statistical test, and subsequent misinterpretation of a P-value.
9   It is important to identify potential sampling or non-sampling errors and to avoid carrying forward these errors when communicating findings from data.

References to support this chapter

BBC (2022) Fake news. https://www.bbc.co.uk/news/topics/cjxv13v27dyt (accessed 11 August 2022).
Church of the Flying Spaghetti Monster (2022) Open letter to the Kansas School Board (2006). https://www.spaghettimonster.org/about/open-letter/ (accessed 11 August 2022).

Greenland, S., Senn, S.J., Rothman, K.J., Carlin, J.B., Poole, C., Goodman, S.N. and Altman, D.G. (2016) Statistical tests, P values, confidence intervals, and power: A guide to misinterpretations. *European Journal of Epidemiology*, 31 (4), 337–50. doi: 10.1007/s10654-016-0149-3

Guilford, G. (2014) There's science behind your inexplicably close relationship with your cat. *Quartz*. https://qz.com/304236/theres-science-behind-your-inexplicably-close-relationship-with-your-cat (accessed 9 June 2022).

Tong, F.S. and Aziz, A.F.A. (2006) How to present research data? *Malaysian Family Physician: The Official Journal of the Academy of Family Physicians of Malaysia*, 1 (2–3), 82–85. http://www.ncbi.nlm.nih.gov/pmc/articles/PMC4453119/

12

STATISTICS COMMUNICATED BADLY

An overview of this chapter

This chapter will build on the previous three, as well as Chapters 4 and 5, highlighting examples of poor statistical analysis, along with inaccurate and inappropriate data communication. This chapter will specifically focus on examples of poorly constructed graphs and tables (building on Chapter 4), as well as tips on how to use data displays correctly, which can assist in the creation of a coherent narrative. You will be guided and encouraged to create your own short reports, based on data that you have collected.

Resources to support this chapter

This chapter draws on a range of sources to help show you how statistics are communicated badly, especially with reference to graphs and tables. In this chapter, we use a range of real-world examples to show you poorly constructed graphs and tables, as well as examples of good data displays. You can build your skills by trying the suggested activities on evaluating data displays from a range of sources, incusing journal articles and media stories.

Resource	Date accessed	Location
Misleading graphs: Real life examples, *Statistics How To*, 2022	30 December 2022	https://www.statisticshowto.com/probability-and-statistics/descriptive-statistics/misleading-graphs/
An observational study on how situational factors influence media multitasking with TV: The role of genres, dayparts, and social viewing, *Media Psychology*, 2014	11 August 2022	https://www.tandfonline.com/doi/full/10.1080/15213269.2013.872038

(Continued)

(Continued)

Resource	Date accessed	Location
Eating habits in the population of the Aeolian Islands: An observational study, *Public Health Nutrition*, 2018	11 August 2022	https://www.cambridge.org/core/journals/public-health-nutrition/article/eating-habits-in-the-population-of-the-aeolian-islands-an-observational-study/1A28F1A49CBCC8F502D15CC30D5948FD
Employers are valuing skills over degrees – here are the highest paying jobs that don't require a college degree, *Women's Wear Daily*, 2022	30 December 2022	https://wwd.com/business-news/human-resources/highest-paying-jobs-1235251697/
Outliers, *Top Drawer Teachers*, 2013	30 December 2022	https://topdrawer.aamt.edu.au/Statistics/Misunderstandings/Misunderstandings-of-averages/Outliers
Identifying outliers, *Top Drawer Teachers*, 2013	30 December 2022	https://topdrawer.aamt.edu.au/Statistics/Misunderstandings/Misunderstandings-of-averages/Identifying-outliers

The websites referred to in all the activities are provided as part of this book's online resources. You can find them at **https://study.sagepub.com/jonesstatsliteracy**.

....

12.1 Inappropriate modelling and inaccurate comparisons

A **statistical model** can provide useful information to aid researchers in identifying relationships between variables. They can also be applied to raw data (which can be sample or population-level data) to aid us in making predictions. In Chapter 5, we looked at a common statistical model used to model data, the normal distribution. We also looked at the assumptions that need to be fulfilled when using such a model, enabling us to be verify whether it accurately represents the data. For example, it would not be appropriate to use the normal distribution to model a categorical nominal variable that has a binary outcome (e.g., Yes/No), since the variable would not fulfil many of the assumptions needed to use the normal distribution. As a reminder these are:

1   The distribution is symmetrical and bell-shaped (i.e., shaped like a bell)
2   It is centred at the mean, μ
3   The mean, mode and median all have the same value
4   The total area under the curve is 1

Looking at data from a categorical nominal variable with a binary outcome (such as Yes/No responses), it is evident that the distribution cannot be symmetrical and bell-shaped. The other assumptions would also be difficult to fulfil.

The modelling of data can be useful in making predictions, for example when look-ing for an association between two variables on a dot plot (Chapter 4). Issues can arise when we utilise a statistical model, make assumptions about the variables we are explor-ing, and the associated data, then expect data obtained in, say, a year's time or two years' time to be the same.

A common example of this, which we see every year, is when the A-level and GCSE results are released in England and Wales. Comparisons are often made, with the expec-tation that all students' grades should roughly fall in line with the normal distribution. This expectation is a reductionist way of looking at a variable (student grades) that has many confounders, which can be affected by many other variables that can lead to big differences from year to year. Each cohort of students that finish their qualifications are different, with different backgrounds, different life experiences, and different socialisa-tion processes that can all have a bearing on the grades they achieve. In addition, the expectation that most students should fall within 2 standard deviations either side of the mean sends out the wrong message to students. It is telling them that there is an expectation that most will never be able to achieve the top grades, even before they have started their respective courses. This is where statistical language, the interpreta-tion of data, and the language we use when looking at data across time, in relation to the statistical models we use, need to be carefully thought through, interpreted and communicated.

········· Develop your skills! 12.1 ····· ··

Figure 12.1   Media 'love' functions
Source: Photo by Karsten Winegeart on Unsplash

(Continued)

Look at the article titled 'An observational study on how situational factors influence media multitasking with TV: The role of genres, dayparts, and social viewing', 2014 (at https://www.tand-fonline.com/doi/full/10.1080/15213269.2013.872038), which you met in Chapter 9, then answer the following questions:

1   Look at the Method section and read the first part titled 'Data' (it is included below, in case you cannot access the online webpage).

> This article is the result of collaboration between academia and industry. The Council for Research Excellence (CRE) (http://www. research excellence.com) is an independent group of research professionals that have conducted a host of studies on media consumption behaviors. For the analysis, we use data from the Video Consumer Mapping (VCM) study conducted from March 26, 2008 to July 24, 2008 .... Nielsen initially provided CRE a list of former participants from their Peoplemeter panel. From this list, 495 U.S. adults were recruited from six designated market areas (DMAs), specifically Dallas, Philadelphia, Atlanta, Seattle, Chicago, and Indianapolis, and were observed for an entire day.

Are there any indications as to the quality of the data used in this study? Explain your answer.

2   Look at the independent variables in the Methods section and pay close attention to the way these variables are described. What types of variables are Genres, Day parts and Social viewing?

3   Look at the dependent variable in the Methods section and pay close attention to the way it is described. What type of variable is Multimedia tasking?

4   The study group used a statistical test called ANCOVA as part of a hypothesis testing approach. Here are two assumptions made when using this statistical test:

A.   Independent variables (minimum of two) should be categorical variables.

B.   The dependent variable should be continuous (measured on an interval scale or ratio scale).

Reflecting on your answers to Questions 2 and 3, are these assumptions met?

5   Part of the results is shown below:

> To test the hypotheses we used a univariate analysis of covariance (ANCOVA) with genre (5), dayparts (3), and social viewing (2) as independent variables and the number of media consumed simultaneously as dependent variable. We used ... Hochberg's GT2 correction to correct for unequal cell sizes and included gender, age, education, income, media ownership, availability, and the time spent on

watching the five different genres as control variables. Overall, the model was significant. Significant effects are reported at the 99% confidence interval (i.e., $p < 0.01$) because of the large number of observations.

Comment on the language used in this part of the results. Are there any issues present? Think about your answer to Question 4 to help you to answer this question.

Try to avoid:

- Selecting an inappropriate model for statistical analyses.
- Ignoring confounding factors when comparing data points across time, and when making predictions with existing data.

12.2 Data presented badly

12.2.1 Scales of measurement

Scales are often used to give the reader a baseline, or a set of reference points, to make a comparison with a data point or response. Back in Chapter 4 we looked at different levels of measurement. For example, we looked at interval and ratio variables, as well as ordinal and nominal. As mentioned in Chapter 4, determining which type of variables you are working with when handling data is an important first step, which can then enable you to decide how to visualise and subsequently analyse the data correctly. There are many examples in journal articles, social media and news stories where the scale of measurement used is incorrect, and this affects how well the data are visualised.

· · · · · Develop your skills! 12.2 · · · ·

This task will enable you to build your ability to evaluate poorly presented data stories. It will also help you to build your statistical literacy confidence.

(Continued)

Figure 12.2 Hiring sign
Source: Photo by Eric Prouzet on Unsplash

Look at the *Women's Wear Daily* webpage titled 'Employers Are Valuing Skills Over Degrees – Here Are the Highest Paying Jobs That Don't Require a College Degree', 2022 (at https://wwd.com/business-news/human-resources/highest-paying-jobs-1235251697/), then answer the following questions:

1   Comment on the presentation of the information displayed on the webpage (focus on the main story, not the pop-ups or adverts). Is it clear? Is any relevant data communicated clearly and consistently? Explain your answer fully.
2   Compare how each job is described, especially with reference to data values.
3   Look at the different variables used in the article, for each job type. Are they explained well? Can you determine what the levels of measurement are for each variable? Are they consistent between jobs?
4   Comment on the statistical language used in the article to explain any data values presented. Is it done well? Explain your answer.
5   Are there any graphs or tables used? If not, do you think the article would benefit from including some? Explain your answer and give several suggestions as to how you would present the data values, using the appropriate data displays.
6   Is the article convincing? Explain your answer.

. . . .

12.2.2 Pie charts and bar graphs

Bar graphs were covered back in Chapter 4, and pie charts have been covered elsewhere in the book. When we reviewed these data displays, we looked at how to visualise data, and the appropriate use of figures and graphs, depending on the types of variables we were working with. When graphs are selected to display data, certain issues can arise, which can lead to misleading or inappropriate ways of reporting and describing the data in them. Sometimes these issues can be due to human error, or a genuine mistake, other times they are deliberate. Remember there is always an agenda behind a newspaper story, news report, social media post or even a journal article. The author is often trying to convince you of their argument or point of view, which can lead them to present data in a way that supports the narrative associated with their angle and, potentially, their perspectives. Common issues can include:

1   The vertical axis has a scale that is too big or too small, or skips numbers, or doesn't start at zero
2   The graph isn't labelled properly
3   Data are left out
4   Percentages don't add up to 100% (in the case of a pie chart)

······· Develop your skills! 12.3 ··· ······························

This task will enable you to build your ability to evaluate poorly presented graphs. This will also help you to build your statistical literacy confidence and skills.

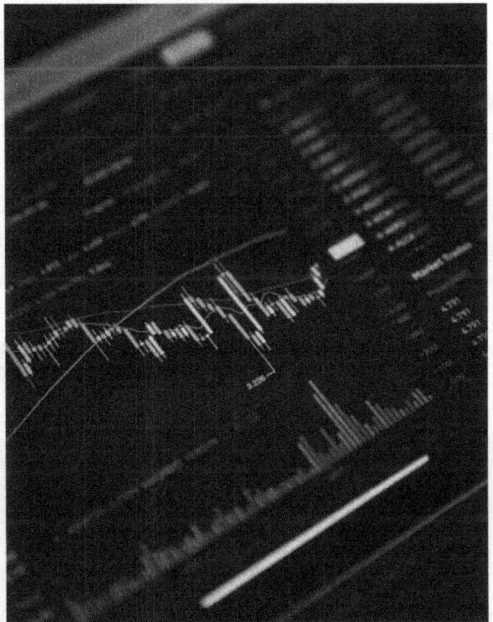

Figure 12.3   Charts
Source: Photo by Dylan Calluy on Unsplash

Look at the Statistics How To webpage titled 'Misleading graphs: Real life examples', 2022 (at https://www.statisticshowto.com/probability-and-statistics/descriptive-statistics/misleading-graphs/), then answer the following questions:

1   Figure 12.4 is taken from the webpage. How is this graph misleading? Explain your answer.

(Continued)

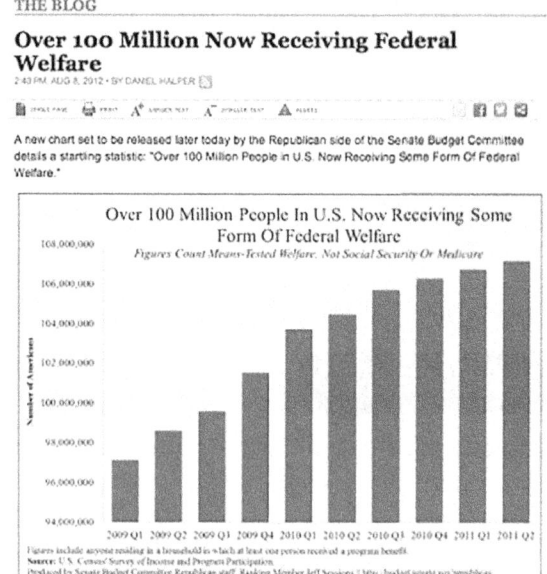

Figure 12.4 Bar graph with a misleading data display

Source: S. Glen, Misleading graphs: Real life examples, *Statistics How To*.

2   Why do you think the author has presented the graph in Question 1, in the way she did?

3   Figure 12.5 is also taken from the webpage. Are there any issues with data display? What's the main problem here?

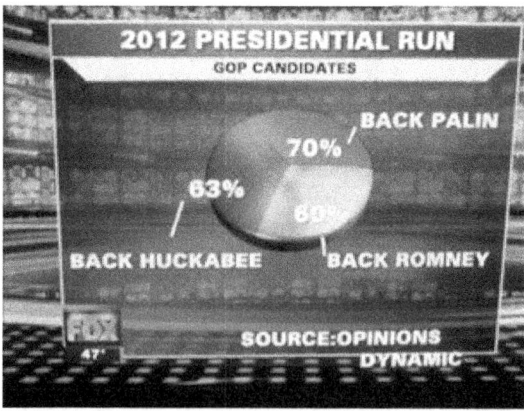

Figure 12.5 Pie chart with a misleading data display

Source: S. Glen, Misleading graphs: Real life examples, *Statistics How To*.

4   How should the data in Figure 12.5 be presented? Which type of graph would be better to display the data presented? Explain your answer.

12.2.3 Outliers

We came across outliers in Chapter 4, when we looked at dot plots and explored their useful properties in being able to help us to identify if an outlier is present in a sample. We also investigated the consequences of outliers, and their impacts on the mean value of a sample. The difficulty in giving advice on how to spot an outlier in a sample of data is to do with the need for contextual knowledge about the data, and the methods used when collecting the data. As you will find out in Develop your skills! 12.4, there is a standard formula used to help to identify an outlier, but what this formula does not factor in is the context behind the numbers, and whether the data point concerned really is an outlier. To do this you need to think carefully about the context and, if possible, gain contextual knowledge about the topic or theme from a subject specialist. The question you'll need to ask is whether the data point in question really is an outlier. It is also useful to ask the researchers or those involved in the data collection process about the method used to collect the data. They should be able to provide insights into whether the data point in question could be due to human error or some other sort of error, or whether it could really be an outlier.

Another strategy for dealing with an outlier is to conduct any relevant statistical analysis both with and without it. This way you can see what effect it has on the outputs and results from the analysis.

· · · · · · Develop your skills! 12.4 ·

This task will enable you to build your ability to evaluate poorly presented data stories. Developing these skills will also help you to build your statistical literacy.

Figure 12.6 Outlier?
Source: Photo by Will Myers on Unsplash

(Continued)

We will look at two webpages from the Top Drawer Teachers website, which is a resource for teachers of mathematics.

1 Look at the webpage titled 'Outliers' (at https://topdrawer.aamt.edu.au/Statistics/Misunderstandings/Misunderstandings-of-averages/Outliers). Comment on the statistical language used. Are there any issues or parts that are unclear? Explain your answer.

2 Now look at the webpage titled 'Identifying outliers' (at https://topdrawer.aamt.edu.au/Statistics/Misunderstandings/Misunderstandings-of-averages/Identifying-outliers). What is the danger in using the formula presented to identify an outlier? Do you think it will always help you to identify an outlier? Explain your answer.

12.2.4 Table formats

We have come across the use of tables at multiple points in this book. Tables are ubiquitous, in fact you have probably seen several tables already at some point today! Remember that tables need to be clear, correctly labelled, with clear column headings (and row headings if appropriate, for example in a two-way tables of counts), and units included if appropriate. They should not be too long or busy to the point that they unclear and confusing.

Develop your skills! 12.5

Figure 12.7   Eating habits

Source: Photo by Katie Smith on Unsplash

Read the first paragraph of the Results section of the article titled 'Eating habits in the population of the Aeolian Islands: An observational study' (at https://www.cambridge.org/core/journals/
public-health-nutrition/article/eating-habits-in-the-population-of-the-aeolian-islands-an-observational-
study/1A28F1A49CBCC8F502D15CC30D5948FD), which you met in Chapters 8 and 9, then answer the
following questions.

Results

Initially, 671 individuals were contacted for the enrolment. Eighty-five denied their
consent or refused to participate. A total of 586 individuals (age range 15–93
years; mean 52 (sd 18) years) accepted to participate in the study; 267 were men
(45·6%) and 319 women (54·4%). Participants' characteristics and stroke risk factors
are shown in Table 12.1; 6·3% suffered from diabetes and 1·27% had a previous
stroke. Almost 40% of the study population was overweight and almost 40% was
normal weight.

Table 12.1 Characteristics and conventional stroke risk factors of the Aeolian population
sample (n = 586), June 2007

Characteristic/risk factor	n	%
Age (years), mean	586	52
SD		18
Sex (male)	267	45.6
Migraine (and headache)	129	22.0
Hypertension	120	20.5
Smoke	84	14.3
Hypercholesterolaemia	79	13.5
Familiarity for stroke	67	11.5
Cardiopathy	43	7.3
Diabetes	37	6.3
Oral contraceptives	19	3.2
Previous stroke	7	1.3
BMI		
Severe thinness (<16.0kg/m²)	1	0.2
Moderate thinness (16.0–16.9kg/m²)	3	0.4
Under weight (17.0–18.5kg/m²)	22	2.6
Normal weight (18.5–24.9kg/m²)	230	39.4
Overweight (25.0–29.9kg/m²)	220	37.6
Moderate obesity (30.0–39.9kg/m²)	108	18.5
Severe obesity (≥40.0kg/m²)	7	1.3

Source: Eating habits in the population of the Aeolian Islands: An observational study, *Public
Health Nutrition*, 22(9), 1590–6, reproduced with permission

(Continued)

1   Comment on the Results section's first paragraph. Is it clear? Explain your answer.
2   Can you think of any way the section you looked at to answer Question 1 could be improved to make it clearer?
3   Comment on the presentation of Table 12.1. Describe any issues with the table in terms of clarity.
4   How could Table 12.1 be improved?

. . . .

12.2.5 Decimal point accuracy

Decimal points are an essential part of data communication. We see them every day on our mobile phones, on the TV, on social media posts and all sorts of other forms of data communication. There are also many examples of bad practice when it comes to the use of decimal points in a data value. For example, using too many decimal places, or including a decimal point where one is not necessary. Errors in using decimal points can make a data story unclear and inconsistent.

· · · · · · Develop your skills! 12.6 ·

Figure 12.8   Healthy eating

Source: Photo by Kelly Vise loin on Unsplash

Now read the first paragraph of the 'Eating habits' subsection of the Results section of the article titled 'Eating habits in the population of the Aeolian Islands: An observational study' (at https://www. cambridge.org/core/journals/public-health-nutrition/article/eating-habits-in-the-population-of-the-aeolian-islands-an-observational-study/1A28F1A49CBCC8F502D15CC30D5948FD), then answer the following questions.

Eating habits

Weekly consumption frequencies of main foods (percentage rates) are summarised in Table 12.2. Percentage rates and 95% CI of deviations from the Italian Institute of Nutrition guidelines are shown in Tables 12.3–12.5. Data from the FFQ evidenced low fish consumption in 13·3% and vitamin intake deficiency in 5·8% of the participants. Excess intake of simple sugars and of animal protein was observed in 8·9 and 5·0%, respectively. We did not observe large differences between males and females except for sodium and calcium intakes. Interestingly, we observed a marked excess of saturated fats (71·0% of people), especially in younger participants. Sodium excess was found almost in half of the participants (49·0%). When considering age groups, we noticed less inappropriate energy intake and less sodium excess in people aged >65 years. Notably, the Aeolian population exhibited a high consumption of olive oil (77·7% used it every day), a very low use of other fats (seed oil, butter, margarine), a scanty use of cured meat (60% never used it) and a low consumption of milk products.

Table 12.2 Weekly consumption frequencies of main foods (percentage rates) in the Aeolian population sample (n = 586), June 2007

	Cereals		Vegetables-legumes-fresh fruit			
	Pasta/ rice	Bread/ crackers	Vegetables	Legumes	Minestrone	Fresh fruit
Never or rarely	10.3	23.8	13.5	13.5	29.1	13.9
1–2 portions	8.9	2.0	16.9	45.7	48.1	5.1
3 portions	10.1	4.1	22.8	23.1	11.5	2.7
4–7 portions	47.3	32.9	39.0	16.1	10.3	32.9
Each meal or almost	23.5	37.2	7.7	1.5	1.0	45.4

	Meat/fish			Milk products		
	Meat	Cured meat	Fish	Cheese	Milk	Yoghurt
Never or rarely	38.4	59.9	15.7	50.0	63.5	81.8
1–2 portions	15.9	18.5	44.5	14.7	6.0	7.4
3 portions	22.1	8.7	18.7	12.3	3.9	4.4
4–7 portions	21.6	12.2	18.8	19.9	20.7	5.5
Each meal or almost	2.0	0.7	2.2	3.1	5.8	0.9

	Eggs	Sweets
Never or rarely	25.4	29.9
1–2 portions	54.7	41.9
3 portions	12.9	9.3

(Continued)

Table 12.2 (Continued)

	Eggs	Sweets
4–7 portions	5.4	15.6
Each meal or almost	1.0	3.4

	Added fats during or after preparation				
	Olive oil	Seed oil	Butter	Margarine	Mayonnaise
Never or rarely	9.8	93.8	97.9	99.3	96.4
Often	12.5	4.8	1.9	0.5	2.4
Always or almost	77.7	1.4	0.2	0.1	1.2

Source: Eating habits in the population of the Aeolian Islands: An observational study, *Public Health Nutrition*, 22(9), 1590–6, reproduced with permission.

Table 12.3 Percentage rates (and 95% confidence interval) of deviations from the Italian Institute of Nutrition, by age group, among the Aeolian population sample (*n* = 586), June 2007

	Age < 46 years			Age 46-65 years			Age >65 years			X^2
	%	95%	CI	%	95%	CI	%	95%	CI	
Excess of saturated fats	80.2	74.2,	85.4	71.7	65.1,	77.6	57.7	49.3,	65.4	22.6**
Inappropriate energy distribution	70.8	64.2,	76.8	71.2	64.6,	77.2	59.6	51.6,	67.2	7.0*
Calcium intake deficiency	56.8	49.8,	63.5	64.1	57.2,	70.6	62.1	54.1,	69.6	2.5
Sodium excess	58.2	51.2,	64.9	55.4	43.5,	57.3	34.7	27.4,	42.6	20.4**
Energy excess	47.4	40.5,	54.3	49.6	42.6,	56.4	43.4	35.6,	51.5	1.3
Fibre deficiency	375	31.0,	44.4	32.5	26.2,	39.3	40.3	32.7,	48.3	2.5
Fats excess	28.1	22.2,	34.7	36.7	30.2,	43.6	29.8	22.8,	37.5	4.0
Low fish intake	14.0	9.7,	19.4	10.8	7.0,	15.8	15.5	10.3,	22.0	1.9
Excess of simple sugars	12.2	8.1,	17.3	5.6	2.9,	9.6	8.7	4.8,	14.1	5.6
Vitamin intake deficiency	6.1	3.2,	10.2	5.6	2.9,	9.6	5.5	2.5,	10.3	0.0
Excess of animal protein	6.5	3.6,	10.7	3.3	1.3,	6.6	4.9	2.1,	9.5	2.4

Source: Eating habits in the population of the Aeolian Islands: An observational study, *Public Health Nutrition*, 22(9), 1590–6, reproduced with permission.

1 Comment on text of the 'Eating habits' subsection in terms of its accuracy and clarity. Pay close attention to the use of decimal points, and comment on the accuracy of their use.
2 Compare the data presented in the written text and the tables presented in this task. Are there any inaccuracies or areas that are unclear? Explain your answer.
3 Comment on the tables presented in this task. Are there any inconsistencies? Are there any aspects of the tables that could be improved? Explain your answers.

Figure 12.9 Data collection and analysis
Source: Photo by Firmbee.com on Unsplash

This task will hone your skills in being able to present a compelling data story. Choose a topic or theme that you are interested in, perhaps a burning question you would like answered. For example, does my pet eat more food at the weekend, compared to a weekday? Then think about how you are going to collect data to answer your research question. When you have constructed your data collection instruments (be careful to think about the ethics of your research question and data collection technique, for example it would be unwise to explore alcoholism in your neighbourhood, which could be unethical and unachievable). When you have collected your data, think about how you are going to present them in an interesting data story. This could be in the form of a poster or presentation, or even a social media post. Remember to reflect on and use all the skills you have picked up throughout this book, especially the generalisability of your findings. Be bold and be brave, present your story to your friends and/or family. Think about the clarity of your story, and the best way to communicate your findings to your audience. Below are some examples you could explore if you are struggling to think of interesting research questions:

1 People eat more chocolate in the evening
2 My phone is spying on me
3 Social media platforms influence my diet
4 Climate change is becoming increasingly common on social media
5 Summers are getting warmer in the UK

Try to avoid:

1 Using the wrong scale of measurement for a variable or set of variables of interest. This can make data presentation and statistical analysis difficult and even incorrect. For example, treating a variable that explores your favourite kind of clothes worn at the weekend as a continuous variable.

2 Presenting pie charts or bar graphs that are poorly labelled, or that have axes that are unclear, and that present data in a misleading way.

3 Assuming a data point is an outlier, without carefully thinking through the topic or context linked to the data.

4 Presenting tables that are unclear or too busy.

5 Using decimal points in an inconsistent or unclear way.

Key points to remember

1 Think carefully about the selection of an appropriate model (e.g., the normal distribution) for the data you need to analyse statistically. Think about the assumptions the data need to fulfil, for you to be confident to use the model of choice.

2 Confounding factors and context are important considerations when comparing data points across time, and they also need to be carefully thought through when making predictions with existing data.

3 Using the correct scale of measurement for a variable or set of variables of interest is extremely important. This step is essential to enable data presentations and statistical analysis to be correct and coherent.

4 Charts and graphs need to be carefully selected and presented in a clear and accurate manner.

5 Be careful when ascertaining whether a data point is an outlier and ensure you investigate the topic or context linked to the data. Also think about the methods used to collect the data, if appropriate thinking about where the data point could be due to human or some other form of error.

6 When using tables as part of a data story ensure they are clear and labelled well.

7 Ensure you use decimal points correctly and consistently in the presentation of data.

References to support this chapter

Fairchild Studio (2022) Employers are valuing skills over degrees – here are the highest paying jobs that don't require a college degree, *Women's Wear Daily*. https://wwd.com/business-news/human-resources/highest-paying-jobs-1235251697/ (accessed 30 December 2022).

Glen, S. (2022) Misleading graphs: Real life examples, *Statistics How To*. https://www.statisticshowto.com/probability-and-statistics/descriptive-statistics/misleading-graphs/ (accessed 30 December 2022).

La Spina, P., Savica, R., Ciacciarelli, A., Cotroneo, M., Dell'Aera, C., Grillo, F. ... Musolino, R. F. (2018) Eating habits in the population of the Aeolian Islands: An observational study. *Public Health Nutrition*, 22 (9), 1590–1596. https://doi.org/10.1017/S1368980018003397

Top Drawer Teachers (2013) Outliers. https://topdrawer.aamt.edu.au/Statistics/Misunderstandings/Misunderstandings-of-averages/Outliers (accessed 30 December 2022).

Top Drawer Teachers (2013) Identifying outliers. https://topdrawer.aamt.edu.au/Statistics/Misunderstandings/Misunderstandings-of-averages/Identifying-outliers (accessed 30 December 2022).

Voorveld, H.A.M. and Viswanathan, V. (2014) An observational study on how situational factors influence media multitasking with TV: The role of genres, dayparts, and social viewing. *Media Psychology*, 18 (4), 499–526. https://doi.org/10.1080/15213269.2013.872038

13

THE POWER OF STATISTICAL LITERACY AND STATISTICAL REASONING

An overview of this chapter

This chapter will review the essential statistical literacy skills covered throughout this book, as well as introducing you to several new skills that will help you in presenting and communicating statistics. This will include encouraging you to present your own data narratives and well-thought-out data displays to produce statistically sound reports. There will also be a focus on seeking feedback on your data stories, as well as the importance of reviewing your own work.

Statistical literacy and statistical reasoning involve the ability to appreciate the art, science and practice within subjects where data (numbers in context) are used, needed or produced. Becoming statistically literate does not happen overnight, and it takes time for the ideas and concepts to sink in, as well as a willingness to appreciate that variability is all around us and impacts everyone. Harnessing your skills in being able to convey an engaging and clear data story will take you extremely far in your studies or place of work, skills that will serve you well for the rest of your life.

Resources to support this chapter

This chapter draws on a range of sources to take you to the next level in relation to statistical literacy and statistical reasoning. You can build your skills by trying the suggested activities for being able to construct your own data stories, as well as responding to feedback from yourself and others.

Resource	Date accessed	Location
How to persuade your audience with data storytelling (+examples), *WordStream*, 2022	15 January 2023	https://www.wordstream.com/blog/ws/2021/05/27/data-storytelling

(Continued)

(Continued)

Resource	Date accessed	Location
Why visual content is important for better user engagement?, *DirectiveGroup*, 2017	15 January 2023	https://www.directivegroup.com/business-strategy-process/why-visual-content-is-important-for-better-user-engagement/
How to tell a story with data: A guide for beginners, *Venngage*, 2021	15 January 2023	https://venngage.com/blog/data-storytelling/

The websites referred to in all the activities are provided as part of this book's online resources. You can find them at **https://study.sagepub.com/jonesstatsliteracy**.

13.1 Bringing it all together

Wow! What a ride we've been on! I'm guessing you had no idea just how many different skills are needed to be statistically literate. And you're probably also thinking that a lot of what you have covered didn't feel like doing mathematics or statistics. I also hope that you have enjoyed yourself along the way! As mentioned in Chapter 2, and across the book in various chapters, statistical literacy is an extremely important and vital element of statistics, and is very different from mathematics and mathematical thinking. The critical thinking and statistical reasoning skills, along with nurturing your ability to develop interesting and achievable research questions that can be answered with good methods that lead to the collection of data that can answer your research questions, have been just some of the skills that we have looked at together. The abilities you have begun to hone can also help you in using data that somebody else has collected, to answer interesting research questions. These skills are not easy to master, and they are not easy to bring together into a coherent and complementary set of statistical tools; however, with perseverance and practice, it is hoped that you will be able to see how valuable they really are, and that you find them useful.

·····Develop your skills! 13.1 ·····

This task will review many of the skills we have covered across this book. It will also encourage you to reflect on work that you have completed in the past, and more recently. The task will also encourage you to think out loud, and to present your ideas based on many of the skills you are beginning to nurture, covered in and supported by this book.

1 Make notes on all the skills you have developed, after completing all the other chapters in this book. Explain the importance of these skills, and think carefully about how you may have used them in other aspects of your life – at university or college, at work, or at other relevant times. It might help you to first talk through your ideas out loud, then make notes at various points.

2 Prepare a presentation (using PowerPoint, or a poster or some other way appropriate way to convey your ideas) describing what statistical literacy is, and why it is so important. You may want to include examples to exemplify your points, and to help bring your presentation to life. Present your ideas to a friend, family member or mentor. Ask them what they think and see if they understand your key points.

Figure 13.1 Skills

Source: Photo by Tim Mossholder on Unsplash

. . . .

Try to avoid:

1 Thinking statistics is the same as mathematics. They are not the same, they are very different, and require different skill sets.
2 Worrying that you need to have mastered all the skills covered in this book, straight after you have covered the relevant content. Learning can be a different journey for many people, and in the case of statistical literacy, it takes time to master all the relevant skills. So be patient!

13.2 Critiquing the work of others and your own

We have critiqued many different forms of data stories at several points in this book, developing key techniques which should enable you to determine what makes an engaging and interesting body of work. We have looked at why people write, the potential reasons for the angle they are taking, and the biases that can be presented with data.

We have also examined the importance of structure and language in being able to communicate key findings from data analysis.

You have probably realised by now that critiquing the work of others is much easier than being able to create your own work. However, since we have looked at the importance of critique, it makes sense to seek critique of your own work. When doing so, it is important to get in the right mindset. Critique is valuable and helps you grow. Many people suffer from impostor syndrome, that feeling that what you have created isn't good enough, which can make critique feel so personal. But if you want to grow personally, and improve your ability to create engaging and interesting data stories, you need to prioritise receiving critique. Finding people you trust, or a mentor, can help with impostor syndrome/sensitivity. When receiving critique, it's also important to fight the temptation to be defensive. Rather than explain yourself, try to lead with 'tell me more'. And try to ask deeper questions about the feedback you have received, such as why it would be better to present your data narrative in a certain way rather than the way you may have presented your data. Even feedback that feels rude and personal, or critique that feels harsh, can be useful, and it would be a shame to lose the chance to improve because you're taking it personally. Do a critique of your own work that you may have created recently, but also of your old work, work that you may have created several years ago. It's good to see where you have come from, and hopefully where you are now, which can help you to evaluate how much you have improved. Developing a habit of critiquing your own work, and work you have created years ago, will not only make you better at critique, but it will also make you more comfortable receiving it.

········ Develop your skills! 13.2 ···· ···

This task will require you to critique a piece of work you have created both recently, and in the past.

1   Select a piece of work that you would class as being a data story, that you have created in the past. This could be a presentation, a poster, or a report. It could have been one you created in school, for work or some other reason. Using the skills you have identified in Develop your skills! 13.1, critique the work you have selected. Comment on the statistical literacy elements of your work, where appropriate, and identify ways in which it can be improved.

2   Select a piece of work that you would class as being a data story, that you have created recently, within the last 6 months. This could be a presentation, a poster, or a report. It could have been one you created in school, for work or some other reason. Using the skills, you have identified in Develop your skills! 13.1, critique the work you have selected. Comment on the statistical literacy elements of your work, where appropriate, and identify ways in which it can be improved.

3   Compare and contrast the work you have selected for Questions 1 and 2. Is there an improvement? Outline the ways in which they are different, and hopefully the ways in which you have improved your abilities in statistical literacy.

Figure 13.2   Critique

Source: Photo by Markus Winkler on Unsplash

....

Try to avoid:

1   Avoiding critique or not showing other people your work. In seeking out feedback and getting others to critique your work, not only will your data stories improve, but you should also become better at critiquing the work of others.
2   Being sensitive or taking things personally when receiving critique on your own work. Remember when people give you a critique of your own data stories, their intentions are most probably to help you improve and get better.

13.3 Writing your own statistical reports and data stories

We have covered many important steps and elements, to help improve your statistical literacy confidence. Building these skills will also enable you to write your own data stories and reports. One of the first steps, which we have looked at in previous chapters in the book, involves thinking about a research question. As a reminder, the following points can help to create interesting and achievable research questions:

1   What is the problem or issue being addresses? Every good research project aims to investigate an existing problem.
2   Who cares about this problem and why?
3   What have others done?
4   What research questions could help to answer the problem identified?
5   What techniques and methods could be used to help answer the research questions?

Being able to answer the research question posed involves selecting the appropriate methods, which can also include a review of the resources needed. This may also require consultation with the relevant subject specialists or people who know the theme of your research area well. Reviewing the relevant and available research literature, based on the theme area, will also help you to interpret the results collected in your study. This will be especially important when you discuss the results collected and make your final conclusions.

When you have collected the relevant data, it's time to think carefully about how to present your findings to help answer your research question. We have reviewed many data stories in this book, as well as looking into the language used in many different examples and formats. We have also looked at different ways to present data, in the form of tables or graphs, and have highlighted pitfalls to avoid. Ensure you think through your data presentation formats carefully, especially in the context of the narrative you are trying to convey, and the path you want your readers to follow.

Remember that the main goal of a compelling and engaging data story, which could be a statistical report, an oral presentation with a supporting digital presentation, or perhaps even a scientific report, is to help guide the reader in a clear and concise manner. You must convince them of your findings and arguments, which could link to existing findings, knowledge and information based on the theme of the data story. This shouldn't mean that you need to resort to fake news or presenting graphs and tables in a misleading or unclear way. Think about the clarity of your story, and the best way to communicate your findings to your audience.

The final step before you release your data story to others is to ensure that you read through your work thoroughly, and check for clarity and conciseness. Think about areas that other readers might find confusing or difficult to interpret or understand. Is there any way you could make it clearer for them to follow? This might seem like common sense; however, you would be surprised how often people forget to read their own work! Make sure you get into the habit of reviewing your own work before you give it to others to read and critique.

·····Develop your skills! 13.3 ·····

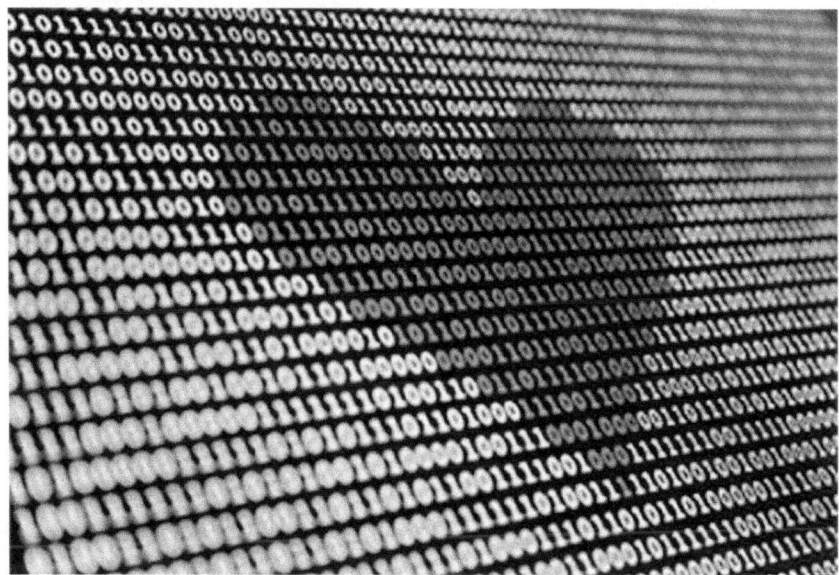

Figure 13.3 Data
Source: Photo by Alexander Sinn on Unsplash

This task will help you to further develop your skills in presenting a compelling data story. Think about a research question you would like answered, and then think about how you are going to collect data to answer your research question. When you have collected your data, think about how you are going to present them into an interesting data story. This could be in the form of a poster or presentation, a written piece, or even a social media post. Remember to reflect on and use all the skills you have picked up throughout this book, especially the generalisability of your findings. Be bold and brave. Present your story to your friends and/or family. Think about the clarity of your story, and the best way to communicate your findings to your audience. Below are some examples you could explore if you are struggling to think of interesting research questions (some are easier to operationalise than others!):

1   Social media influences what I feed my pet
2   The Metaverse will be the future of social activities for the majority of people
3   Comedies are becoming increasingly popular on streaming services
4   Cat owners have higher levels of self-esteem
5   Eating less meat is healthier for you

Try to avoid:

1   Creating research questions that are unachievable, too broad, or unethical.
2   Presenting data in a misleading or unclear manner.

13.4 Checklist to build your confidence with statistical literacy

This chapter concludes with a checklist to help build your confidence with statistical literacy. Hopefully these points will complement and indeed overlap the skills you are beginning to nurture, and will be useful as a reminder.

1   Statistical methods should enable the use of data to answer scientific questions: Ask 'why am I doing this?', rather than focusing on which technique to use.
2   You can never eliminate chance as a factor when making conclusions, especially with experiment-to-causation inferences.
3   Worry about data quality, and interrogate the methods used to collect them (if this is possible), as well as potential biases. Everything rests on the data.
4   Statistical analysis is more than a set of computations. Do not just blindly plug numbers into formula or run procedures in software.
5   Keep it clear and concise. The main communication from a statistical analysis should be as clear as possible. It should also weave nicely into a data story, or report.
6   Check the assumptions underlying any analyses you carry out. Make clear any reservations you may have as to whether these have been satisfied. Make your analysis reproducible by allowing the data you have used to be freely available and accessible. Others should be able to access your data to help them interpret your data displays and data stories.

········ Develop your skills! 13.4 ···· ································

This task involves looking at several webpages that offer tips on how to create good data stories, using essential statistical literacy attributes. Evaluate the following webpages and explore whether they offer any new and useful skills. Then answer the questions below:

a   How to persuade your audience with data storytelling (+examples), 2022 (at https://www. wordstream.com/blog/ws/2021/05/27/data-storytelling)
b   Why visual content is important for better user engagement?, 2017 (at https://www. directivegroup.com/business-strategy-process/why-visual-content-is-important-for-better-user-engagement/)
c   How to tell a story with data: A guide for beginners, 2021 (at https://venngage.com/blog/data-storytelling/)

1   Is there anything that surprises you after reviewing each of the webpages?
2   Can you spot any biases in reading the webpages? Think about the intended audience of the webpages to help you answer this question.
3   Are the webpages convincing? Explain your answer fully.

Figure 13.4   The passion for data storytelling
Source: Photo by Ian Schneider on Unsplash

Key points to remember

- Statistics is not the same as mathematics. They are very different, and require different skill sets.
- Try not to worry that you need to have mastered all the skills covered in this book, so soon after you have covered the relevant content. Learning is often a different journey for many people, and in the case of statistical literacy, takes time to master all the relevant skills. So be patient!
- Critiquing your own work, and showing other people your work, is very healthy and can give you useful tips and ideas to improve. In seeking out feedback and getting others to critique your work, not only will your data stories improve, but you should also become better at critiquing the work of others.
- Try not to be overly sensitive or take things personally when receiving critique on your own work. Remember when people give you a critique of your own data stories, their intentions are most probably to help you improve and get better.
- When creating research questions, make sure that they are achievable, specific and ethical.
- Present your data in a clear and concise manner that doesn't mislead. This includes carefully selecting the appropriate data displays, for any data you are aiming to include in your data narratives.

References to support this chapter

Hooper, L. (2021) How to tell a story with data: A guide for beginners, *Venngage*. https://venngage.com/blog/data-storytelling/ (accessed 15 January 2023).

Maier, L. (2017) Why visual content is important for better user engagement?, *DirectiveGroup*. https://www.directivegroup.com/business-strategy-process/why-visual-content-is-important-for-better-user-engagement/ (accessed 15 January 2023).

WordStream (2022) How to persuade your audience with data (+examples). https://www.wordstream.com/blog/ws/2021/05/27/data-storytelling (accessed 15 January 2023).

APPENDIX

ANSWERS TO CHAPTER QUESTIONS

The suggested answers provided below are presented as a guide, linked to the 'Develop your skills' questions throughout the book. Your own answers may be different to the ones presented in this chapter.

Chapter 2

Develop your skills! 2.1

1   There were lots and lots and lots of eggshells all over the very big floor in the kitchen by the toilet, across the way from the living room.

Rewrite as: There were a lot of eggshells on the large kitchen floor. The kitchen was by the toilet, across from the living room.

Additional guidance/explanation. The original sentence had too much information in it, so was split into two more concise and clear sentences. The phrase 'lots and lots and lots' was also replaced with 'a lot', which reads better. The 'very big floor in the kitchen' was rearranged to 'large kitchen floor', which is more concise and clearer.

2   The most favourite sandwich in Burger King is a Whopper, then it's a chicken royale, then it's a cheeseburger.

Rewrite as: Customers voted on their favourite sandwich in Burger King, with the Whopper having the highest proportion of votes, followed by the chicken royale and then the cheeseburger.

Additional guidance/explanation. The original sentence has been rewritten to be longer, but now it reads a lot better and is easier to understand. For example, the 'most favourite sandwich' has been replaced with part of the sentence giving a little more information on where the votes have come from (customers). The various sandwiches from Burger King are also explained better, describing which one had the highest proportion.

3   Most people seem to very much like to at 9 a.m., run for, on average, about 34.5 minutes, then have breakfast after, in many suburbs in Auckland, which is in New Zealand, on the North Island.

Rewrite as: The highest proportion of people run at 9 a.m. for an average of 34.5 minutes, in several suburbs in Auckland. They then have breakfast after running.

Additional guidance/explanation. The original sentence was poorly structured and difficult to follow. To help improve it, the sentence was split into two to make it more concise. Information was also added to give the reader an idea of what was being measured: the proportion of runners. Some of the extra detail from the original sentence was also dropped, since most people should know that Auckland is in New Zeeland, and there really is no need to include the North Island.

4   Most people with a virus who contract a virus and get ill from it will recover, in many countries across the globe.

Rewrite as: The majority of people who contract a virus and fall ill will recover, on a global scale.

Additional guidance/explanation. The original sentence was changed to drop the repetition of the word 'virus'.

5   The results from an observational study proved that 54% of people lost so much of their concentration in the workplace, mostly males were affected, when they watched a lot of TV, which is about 5 hours or more of their spare time.

Rewrite as: The results from an observation study suggest that 54% of people who watch TV for more than 5 hours of their spare time a day often lost concentration in their workplace. The study also reported that mostly males were affected.

Additional guidance/explanation. Observational studies often produce results that suggest causality but cannot prove a link, which is why 'suggest' has been used instead. This will be covered more in later chapters. The sentence has been tidied up and split into two to ensure the information flows better.

Develop your skills! 2.2

1   People from lower socioeconomic groups are more likely to lose their jobs and are
     more likely to use their savings to survive. During times of economic stress,
     people from different socioeconomic groups, in Western society, react differently.
     People in the Middle East will react differently. Whereas people from higher
     socioeconomic groups are more likely to invest their savings and top up their
     pensions. In conclusion, the rich get richer, whereas the poor get poorer.

Rewrite as: During times of economic stress, people from different socioeconomic groups,
in Western society, react differently. People from lower socioeconomic groups are more
likely to lose their jobs and are more likely to use their savings to survive, whereas people from higher socioeconomic groups are more likely to invest their savings and top up
their pensions. In conclusion, the rich get richer, whereas the poor get poorer.

Additional guidance/explanation. The sentences have been arranged to form a logical and
coherent flow of information. The sentence referring to people in the Middle East has
been removed, since it should form part of another sentence.

2   Evidence-based approaches should be used instead. Learning involves a
     combination of cognitive abilities and life experiences. Many learning theories are
     contentious, and some are underpinned by weak empirical evidence. A more
     nuanced approach in using learning theories in relation to teaching practices,
     such as Bloom's taxonomy, should be adopted. Despite this, educators often take
     learning theories and use them as if they are facts, embedding them in the
     teaching methods and guidance developed.

Rewrite as: Learning involves a combination of cognitive abilities and life experiences.
Many learning theories are contentious, and some are underpinned by weak empirical
evidence. Despite this, educators often take learning theories and use them as if they
are facts, embedding them in the teaching methods and guidance developed. A more
nuanced approach in using learning theories in relation to teaching practices, such as
Bloom's taxonomy, should be adopted. Evidence-based approaches should be used.

Additional guidance/explanation. This task involves complex words and ideas, which can
prove tricky when trying to present the information in a logical flow. Think about the
path you think would make the most sense to readers.

3   Sadly, the field is also littered with unsubstantiated claims that have little or no
     evidence. Nutritional biochemistry can tell us useful things about the food and
     drink we consume. An interdisciplinary approach continues to aid and contribute
     to the important evidence base of this discipline. As a disciplinary field, the

evidence used to support the knowledge base has been increasingly steadily over the last few decades.

Rewrite as: Nutritional biochemistry can tell us useful things about the food and drink we consume. As a disciplinary field, the evidence used to support the knowledge base has been increasingly steadily over the last few decades. An interdisciplinary approach continues to aid and contribute to the important evidence base of this discipline. Sadly, the field is also littered with unsubstantiated claims that have little or no evidence.

Additional guidance/explanation. This task involves complex words and ideas, which can prove tricky when trying to present the information in a logical flow. Think about the path that would make the most sense to readers.

4   Communicating verbally involves several key aspects that involve several human senses. For example, people's concentration span has decreased over the last decade, as well as their ability to verbally communicate with others. Social media increasingly influences the way many people live their lives. Many scientists claim that this is having a detrimental effect on human behaviour. Data shows that people are spending a significant amount of time using social media platforms on their smart phones or other digital devices. Proponents of social media state that there are many benefits for people engaging with the multiple platforms available, including the creation of new and exciting forms of communication.

Rewrite as: Social media increasingly influences the way many people live their lives. Data shows that people are spending a significant amount of time using social media platforms on their smart phones or other digital devices. Many scientists claim that this is having a detrimental effect on human behaviour. For example, people's concentration span has decreased over the last decade, as well as their ability to verbally communicate with others. Proponents of social media state that there are many benefits for people engaging with the multiple platforms available, including the creation of new and exciting forms of communication.

Additional guidance/explanation. The sentences have been arranged to form a logical and coherent flow of information. The sentence referring to communicating verbally has been removed, since it should form part of another paragraph.

Develop your skills! 2.3

1   The temperature reached 9 degrees by noon.
2   Our afternoon meal cost a total of £55.62.
3   We often liked to travel along the M4, which took us to Oxford.

4   There are approximately 23 billion results when you search for the word 'number' in Google. The search takes 0.7 seconds.

5   It started snowing on 9 February and continued for 35 days.

Develop your skills! 2.4

1   87% of British Facebook users have admitted to looking at their ex-partner's profile, within two weeks of the break-up. 69% of these have also stated that they were the ones who were dumped by their ex-partner.

Rewrite as: In the UK, 87% of British Facebook users have admitted to looking at their ex-partner's profile within 2 weeks of the break-up. Of these, 69% of these have also stated that they were the ones who were dumped by their ex-partner.

Additional guidance. An even more informative example would be to include the sample size, so the reader has an idea of the scale of people who responded. For example, this could be "87% (5500) of British Facebook users…". No sample size is strictly needed the second time a percentage is mentioned, as it can be worked out (69% of 5500), but it could be included at the writer's discretion.

2   The most frequent day for people in the UK to order a takeaway is Friday, a recent poll by Readers Eating suggests. More than 10,000 people responded to the poll, mostly situated in the capital, London. Around 67% said Friday was their most popular day to order takeout, with Monday being the least likely day.

Rewrite as: Friday appears to be the most popular day for Londoners to order takeaway, according to 67% of over 10,000 people who responded in a recent Readers Eating poll. Monday seems to be the least likely day to order takeout. Most of the respondents were from London, but some were from other parts of the UK.

Additional guidance. The sample size has been added to give the reader an idea of how many people were involved. The original sentence was a little repetitive, so the percentage has been included in the first sentence. The target population was changed to Londoners, since most of these people responded to the poll.

3   More than 52% of British adults have stated that they are worried about ageing. Of these, 62% were female, 22% were male and the rest were non-binary.

Rewrite as: Worried about ageing? In a recent poll, 52% have stated that they are worried about the process. Of these, 62% were female, 22% were male and the rest were non-binary.

Additional guidance: 'Worried about ageing?' was added to make it a more eye-catching read.

4   Approximately 7 out of 10 people in Grimsby use music platforms to stream or download music, a new study has found. The most popular platforms used were Spotify and Apple Music.

Rewrite as: A new study has found that 7 out of 10 people in Grimsby use music platforms to stream or download music. The most popular platforms used were Spotify and Apple Music.

Additional guidance. As a rule, try to avoid using words like 'approximately' or 'roughly' or 'more than' for proportions when you have specific values. If after calculating the percentage there is an answer with a decimal, do some rounding and use that value.

Develop your skills! 2.5

1   The application makes several claims, to assist with peoples writing skills: great writing simplified, with the ability to compose bold, clear, mistake-free writing, with the assistance of the new AI-powered desktop Windows app (Grammarly). From grammar and spelling to style and tone, Grammarly's suggestions are comprehensive, helping users to communicate effectively. They also claim that everyone can gain confidence with their writing. They also state that users can get started for free, with the assumption that costs will be incurred at some point. There are also tabs along the top of the home page, which go into more detail for specific types of users, including colleges and university students and users who may need it for work.
2   The application's producers are likely to assume users have a certain level of writing skills, at the very least a minimum level to be able to construct sentences that can be corrected. Of course, within that there is going to be a level of variability, with some users having a higher level of writing skills than others.
3   Users may become over-reliant on the application, which could ironically lead to a deterioration in their writing skills.
4   Prolonged use of the application may lead to a deterioration in writing skills and could impact an individual's ability to spell correctly. Skills in using grammar correctly may also be affected. The ongoing costs could also become expensive.

Chapter 3

Develop your skills! 3.1

This exercise requires you to search for your own surveys of interest and take notes.

Develop your skills! 3.2

This exercise requires you to take notes.

Develop your skills! 3.3

1   The language used in the survey is clear and concise. The questions are positioned to the left of the webpage, in large bold text, with the options for the user to select on the right. The survey uses a five-point Likert scale for users to register their response for each question. Questions are presented one at a time, and users can progress forward to the next question, or go back to the previous one.

2   There is a progress bar, to give users an indication as to how many more questions they need to fill in to complete the survey. I addition, there is a 'save my progress' option, so users can leave the survey and come back to it later. The design is simple, populated with text and a plain colour scheme.

3   Users are likely to be searching for a change of career, or for help to identify an appropriate career pathway, linked to their needs.

Develop your skills! 3.4

1   The survey is clear and uses language most literate individuals can understand. However, sometimes the user might wonder what the relevance is for some of the answer options to the questions, in relation to perfume preference.

2   The survey is populated with many pictures and interactive elements, to grab and maintain the attention of the user. Some of the pictures are comical, which could be the intent of the creators of the survey, to assist in engaging the user. The survey is easy to navigate, culminating in options being presented as to the most suitable perfumes available for purchase, linked to the responses given in the survey. (You do need to provide your email address to receive answers to your responses.)

3   Perfume lovers, people looking for an alternative gift that isn't mainstream or widely available.

Develop your skills! 3.5

1   False
2   True
3   True

Develop your skills! 3.6

1   Non-response and self-selection bias. Potential behavioural considerations as well.
2   The questions mentioned in the scenario were 'Roughly how many hours do you spend a week watching Netflix?' and 'How many people in your household use Netflix?'
    The first question is a little ambiguous: what does 'roughly' mean?
    The second question is better.
3   True
4   True
5   True
6   False

Develop your skills! 3.7

1   Non-response and self-selection bias. Potential behavioural considerations as well.
2   The questions mentioned in the scenario were 'Roughly how often do you shop on Amazon?' and 'How many people in your household use Amazon?'
    The first question is a little ambiguous: what does 'roughly' mean?
    The second question is better and straightforward to answer.
3   False
4   False
5   False
6   True

Develop your skills! 3.8

1   Non-response, self-selection, and interviewer bias. Potential behavioural considerations as well.
2   The questions mentioned in the scenario were 'How many hours do you spend in a petrol station each month?' and 'Are you a member of the rewards scheme offered at the petrol station?'
    The first question is presented well and is specific. The second question is also clear and concise.

Chapter 4

Develop your skills! 4.1

1   A Google search in 2021 would have found that there were 4983 banks. The point
    of this question is to get you to think about questioning numbers in their
    context, and what other knowledge you might need to answer questions like this.
2   Commercial banks
3   Categorical – nominal
4   Mean = $221.5 billion, SD = $515.6 billion
5   Median = $52 billion
6   Looking at the plot, there appears to be several extreme data points, far away
    from the bulk of the data, which is skewing the mean value. These extreme data
    points will have no effect on the median value.

Develop your skills! 4.2

1   Bar chart
2   Dot plot or box-and-whisker plot

Develop your skills! 4.3

1   Histograms display the relative density of observations (i.e., give a good idea of
    the shape of the distribution) and they are good for large amounts of data.

Develop your skills! 4.4

1   Yes (in general), as the x-axis increases, so does the y-axis.

2   *Trend*. There does appear to be a linear relationship – you can roughly fit all the
    dots onto a straight line.

    Scatter. There is some degree of scatter, especially in the bottom left part of the
    plot.

    Strength of relationship. There is a moderately strong relationship between with the
    variables – if you were to draw a straight line through the data points, most of
    them would be close to that line.

    Association. The two variables are positively associated.

3   This will depend on your original answer and whether, after exploring the
    plot further, you stick to your original decision, or change your mind after
    reflection.

Develop your skills! 4.5

1   No, the data do not appear to have three prominent modal peaks.
2   The sample sizes between males and females are clearly different (more males
    than females), though the median values appear to be similar. There could be
    several reasons for this, for example the sampling method used (they may have
    sampled males and females who work in similar companies, with similar jobs,
    that pay similar wages for both groups).
3   Potentially. However, there is no way of knowing this until we see the data
    and the resultant distribution for the larger sample taken for the
    female group.

Develop your skills! 4.6

1

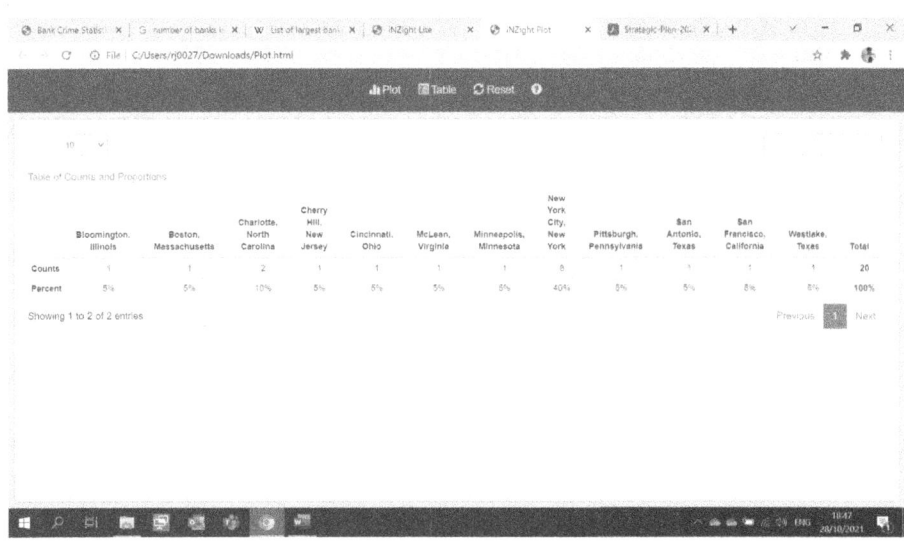

2   New York

Chapter 5

Develop your skills! 5.1

1   Difference or change in the global mean surface temperature (in °C)
2   Carbon dioxide emissions
3   Vertical growth of plants in Milford Sound, New Zealand
4   Likeability
5   Mean time it takes for stars to explode in the universe

6   Financial deprivation in the UK
7   Mean spend online per year

Develop your skills! 5.2

1   Confidence interval for the change in global mean surface temperatures over last
     25,000 years.
2   Because of the scale used in the graph, the section of the last 1000 years needs to
     be enlarged because it is not clear and there is some overprinting (i.e. bunching)
     of data that are too close together.
3   Yes.

Develop your skills! 5.3

1    $8.11 \times 2 = 16.22$
2    $83.6 - 8.11 = 75.49$; $83.6 + 8.11 = 91.71$. Thus we have lower and upper limits of
     (75.49, 91.71).
3   No, since this is a 95% confidence interval. This interval is a range of plausible
     values for the true population parameter, so we can say it is a fairly safe bet that it
     is contained in here.
4   This might be tricky to do since median values are not always in the centre of our
     distribution of data. So unlike norm-based confidence limits, we would not be able
     to assume the media is in the middle of the bootstrap confidence limits. So the
     answer is no!
5    $(6.9 - 2.2)/2 = 2.35$
6   They have misunderstood the type of inference a confidence interval in this
     context can apply to. A confidence interval is a range of plausible values for the
     true population parameter. A confidence interval should not be interpreted or
     extrapolated to other experiments.
7   Yes, it is unfair and generalises with no evidence that all scientists cannot grasp
     what a confidence interval represents.

Develop your skills! 5.4

1   Emotional reactions
2   H_0: There will be no difference in the baby's emotional responses when he is
     shown the white rat.

     H_1: There will be a difference or change in the baby's emotional responses when
     he is shown the white rat.

3   Yes, many! It is unethical to cause mental and/or physical harm to humans – the
     baby might be scarred for life.

Develop your skills! 5.5

1   The *P*-values are presented in an order ranging from smallest (i.e., from highly
    significant to non-significant).
2   They are different and use different phrases to explain what the *P*-values
    represent. When you come onto Section 5.6, you will see that the word
    'significant', as reported in the article for this question, is too general, and
    although it is implied that it is referring to statistical significance, this is not
    stated explicitly in the abstract. Practical significance, which refers to a confidence
    interval, will be discussed in more detail in Section 5.6.
3   Articles like this one are important since they give an indication as to potential
    biases or motivations behind what types of medical research are published and
    made available to the scientific community.
4   Significant results (which we will assume to refer to statistical significance) can
    help to support the claim that there is an association or link between variables in
    an experiment or study. Depending on the type of study undertaken, for example
    a well-designed and well-executed experiment, causal links can potentially be
    made. This is the goal of all researchers across many fields, which could explain
    the higher proportion of significant results being reported by the authors of the
    article. It is often a lot harder to get statistically non-significant results (i.e., where
    there is insufficient evidence against the null hypothesis) accepted for publication
    in a journal, since many scientists may incorrectly interpret this as no causal link
    or association between variables. A statistically non-significant result means you
    cannot rule out chance acting alone, that is, you cannot rule out that the results
    you see in your study are luck of the draw results merely due to chance.

Develop your skills! 5.6

1   Yes, there seem to be confusions over what a result from a *P*-value enables you to claim
    or say. When they refer to extrapolation and generalising to populations, they are
    referring to inferential statements that are more to do with the type of study that is
    undertaken. They have also failed to talk about difference types of inference you can
    make (sample-to-population and experiment-to-causation inference).
2   The main issues stated involved the misuse of *P*-values. The first issue stated
    reports that using the *P*-value alone to conclude whether there is an effect or not
    is a bad practice for interpreting the results, and yet this practice is widely
    prevalent. The second issue describes that '[o]ne analysis using 791 published
    research papers published in 5 different journals found that about half of them
    interpreted the p-values incorrectly by assuming non-significance means no
    effect[;] absence of evidence is not evidence of absence'.
3   Potentially practical significance (i.e., using the confidence interval). They could
    also be referring to effect size, which is not covered in this book.

Develop your skills! 5.7

1   All of them.
2   Experiment 3 was the only experiment where the confidence interval was above 20 seconds for both lower and upper limits.
3   As alcohol levels increase, the mean time it takes to complete the concentration test (i.e., concentration of the participants in the alcohol group) increases.
4   Yes, since a norm-based method was undertaken, which assumes the data from the experiment (for concentration time) are normally distributed. Therefore, we could use the centre of the confidence intervals to find the mean time difference values, from each experiment, between the groups.
5   Yes, there would have to be ethical checks and clearance obtained, since mental and physical harm could present in individuals who become drunk from the study.

Chapter 6

Develop your skills! 6.1

1   The answer to this question will be personal to you and depend on your own beliefs and the ways in which you trust information and the knowledge you have accumulated throughout your life. Pretty heavy philosophical stuff, right? Interesting though!
2   The four steps of the scientific method are:

- Observation – of some measurable aspect of the universe
- Hypothesis – about a property of the universe, based on observation
- Prediction – of something that should hold true if the hypothesis is correct
- Experiment – to test the prediction

The steps can be repeated indefinitely, and any repetition will either support, refute or modify the existing theory.

Develop your skills! 6.2

1   If bacterial growth is related to temperature, then increasing or decreasing temperature may alter bacterial growth.
2   If chocolate consumption is related to spots forming in humans, then changing the amount of chocolate a human consumes could lead to an increase or decrease in spot formation.
3   If educational attainment is related to ethnicity, then we would expect to see differences in educational attainment from different ethnic groups.
4   If an individual's lifespan is associated with diet, then we would expect to see differences in lifespan among people who adopted different diets.

Develop your skills! 6.3

1   Perhaps academics in the field of music might find this research interesting, as it might be of use in their own research. Even people involved in choirs could find it of value, having said which, they are unlikely to look at such journal articles, which are not often written for lay people, assuming the choir members are not academics in the field of music, or other relevant fields.
2   Not entirely clear – the purpose is outlined but it's not posed as a clear research question.
3   This piece is much clearer and easier to understand. It is longer and explains what the research group investigated, in relation to singing accuracy.
4   The research questions are still not clear, but you could take a guess at what they might be, for example: what factors influence singing accuracy? Or are age and singing accuracy related or associated?

Develop your skills! 6.4

1   The data displays are embedded very well within the dynamic webpage. They provide an engaging and integrated addition to the narrative of the story being presented. The way the data displays are presented is understandable, clear and dynamic.
2   The data included within the dynamic data displays report events that happened over time, movements of Jews in and out of Lebanon, as well as visual representations of places and artefacts. The data could have come from official government archives, or from news reports made at the time. The data could have been collected as part of government census data, although population-level census taking doesn't seem to take place in Lebanon, due to the concern over the data collected, and the potential for them to cause civil unrest (from the reporting of the religious status of individuals). Data could have been collected by news agencies to help with the reporting of a relevant news story.
3   Stories based on religion, and in particular the persecution of the Jews, can be highly emotive for some. For these reasons, this data story has been labelled as 'sensitive'.
4   Overall, the webpage is presented very well, with a coherent narrative and embedded engaging data displays. There do seem to be a lot of images that break up the flow of the story somewhat, which I'm sure could be reduced without affecting the impact of the story.

Develop your skills! 6.5

1   The data displays are embedded very well within the dynamic webpage. They provide an engaging and integrated addition to the narrative of the story being presented. The way the data displays are presented is understandable, clear and dynamic.

2   The data included within the dynamic data displays report events that happened
     over time, relating to the major events that led to Brexit, as well as visual
     representations of the key players involved with this historic event. The data
     could have come from official government archives, or news reports made at the
     time. They could have been collected from freely available government/
     administrative data. They could have been collected by news agencies to help
     with the reporting of a relevant news story.
3   Overall, the webpage is presented very well, with a coherent narrative, and
     embedded engaging data displays. There are a lot of images embedded within the
     data story, although these are mostly people involved with the Brexit process. The
     bar graph roughly a third of the way down the page requires the user to hover over
     segments of the bar chunks to reveal what they present. This isn't entirely clear, and
     the graph would benefit from having axes that are labelled to improve its clarity.
     Another bar graph roughly two-thirds of the way down the page has the same issue.

Develop your skills! 6.6

1   The webpage is well laid out, with pop-ups to catch your attention. The content is
     basic, and a little short on text. There are also tabs at the top of the webpage,
     directed at specific users.
2   Overall, the webpage is well presented, basic, but should be convincing enough
     for people to at least give it a try (the website claims it's free).
3   The home page is a little light on content and has large gaps across the page.

Develop your skills! 6.7

1   The webpage is well laid out, with large font text, clear data displays, and
     embedded video content, to grab the attention of the reader. There are also links
     to other news stories embedded in this story.
2   The author of the article most likely sourced the data (looked for the relevant data
     sets linked to the topic) needed for the story to be factually correct. Much of the
     data collected is likely to come from publicly available sources and other news
     articles (either on their own webpage, or externally from other sources).
3   The news story is a good length, and short enough for people to read. They could
     put an indication at the top of how it takes to read. They could also look to
     remove the volume of adverts on the webpage, which are distracting.

Develop your skills! 6.8

1   The webpage is well presented, with a lot of content. There are relevant images and
     video content embedded, as well as an FAQ section, to engage the reader. There is also
     a vertical tab section on the top left-hand part of the page, to aid in easy navigation,

as well as tabs at the top to look at other articles and content. Specific data are included in multiple parts of the webpage, outlining approximate numbers of individuals who fall into each category of influencer stated. There are several interesting parts included on the webpage, including topics such as 'Do influencers pay for followers'? For example, the webpage states that many large celebrity accounts have their share of fake followers, created by bots without the celebrities knowing.

2   The article was created by Werner Geyser, fascinated by the world of online marketing, with 18 years in the fields of social media, ecommerce, email, and influencer marketing industries. The webpage was updated in 2023. The article was most likely written, with associated data and data displays, to enlighten readers on a topic that is currently very popular.

3   The article has in large brackets (in the title) that it was updated in 2023, then underneath it specifically mentions it was updated on 'March 24th, 2023'. One of these could be dropped (I'd suggest deleting the update note in brackets in the title). The sound quality of the embedded video could also be improved.

Develop your skills! 6.9

1   The news story is well laid out, with text that is easy to read, along with relevant images and links to other news stories, on Sky News. With a story of this nature, it contains multiple forms of data, including percentages, quantities, proportions, and monetary values.

2   The author of the news story is Megan Baynes, who also includes a link to her Twitter feed. The opening sentence to the report mentions the Office for National Statistics, as showing a rise in the percentage of the population struggling amid the cost-of-living crisis.

3   The webpage contains too many adverts, which distract from the news story. In multiple parts of the news story, the journalist quotes both a proportion and percentage, which is not always necessary – for example, around one in 15 (7%) of disabled adults reported being behind on their energy bills, compared to one in 25 (4%) non-disabled people. Most people tend to understand percentages well, so the journalist should be consistent and stick to one or the other. Sample sizes are also useful, when quoting percentages, to give the reader an indication as to how many people, or units, the percentage is referring to.

Chapter 7

Develop your skills! 7.1

1   You might be surprised that the article found overwhelming evidence on the positive impact of social media as a language learning and teaching environment. Most experiences people tend to have of using social media,

especially when accessing them on mobile phones in a high school setting, tend to be forbidden.

2   This answer will be specific to your own individual experiences.
3   Again, this answer will be down to your own creativity!
4   Hopefully they liked it and it helped them to better understand a statistical concept or idea you were trying to get across.

Develop your skills! 7.2

1   You can agree or disagree with this article. And you should have supporting statements, potentially with experts from the article, to support you answer.
2   Overall yes, the article does make a strong case for data literacy. The article outlines what data literacy is, why it is important, and presents some statistics on the numbers of UK workers that are statistical literate (not sure what UK workers means though?).
3   Overall they are a good start, to enable readers to critique an article. There are several areas that could be expanded upon or better defined. For example:

    i   In point 1, do people know what validity means?
    ii   Point 2 suggests ditching the news sites and outlets that you know to be sensationalist or false. But how would people know this? Why would you subscribe to sites that are false in the first place, unless you were unaware that they were false?
    iii Point 3 is not entirely clear. What does cleansing social media timelines mean? Is it deleting the browser history? Deleting cookies?
    iv   Point 4 is not clear on what it is asking you to do. Create your own data story? Hopefully the guidance in this book is clearer than this!

Develop your skills! 7.3

1   The article/webpage is presented with a series of subheadings, each followed by several bullet points that report percentages on mostly categorical variables. The webpage is clear and concise, but could benefit from using graphs and plots to better illustrate the data display.
2   People may look at where the information has come from, to see whether it is a trusted site or organisation (e.g., the NHS or BBC). People may also question the actual information itself and the plausibility of the statements and content.
3   Yes, this is a large percentage of coverage from the population of online UK users.
4   The BBC is a well-known and one of the oldest publicly funded organisations in television, which has built up a respectable reputation over the years. It is trusted by many across the world.
5   Younger people might be savvier with data displays and using mobile or laptop devices. They are more likely to access and use online news and social media

platforms and therefore could be better equipped (i.e., more experienced) to spot fake news stories and articles.

6   Behavioural considerations (not wanting to report that they have not been following official advice) is the most likely form of non-sampling bias to explain these results. Non-response bias is also a possibility.

7   The data display will typically include (this will be dependent on which coloured tile you select) a series of drop-down menus on the right (with variables like Age, and levels of measurements being contained within the drop-down menus), questions from the survey towards the top, and then a visualisation of the response underneath the survey questions. There is a side-by-side bar graph in the data visualisation panel underneath the questions from the survey.

8   About once a month.

Develop your skills! 7.4

1   Enjoy reading the article you chose!

2   This answer will depend on the news story/article you chose to read. Most the stories are well presented, and are broken up with subheadings and the use of different font sizes and fonts to keep your attention. They are also short, so you do not have to worry about spending too much time reading them.

3   Again, this will depend on what you click on to read. Most the articles summarise big news stories that are likely to feature fake news or disinformation/ misinformation circulating on social media platforms and online. There are also links to fake news stories, which the BBC tries to investigate and provide the reader with the truth.

4   The news stories are based on the BBC news website, which is a well-respected, long-established and trustworthy organisation. There is also embedded video footage of certain events, linked to the news stories, that have been used to refute fake news.

Develop your skills! 7.5

1   The ASOS webpage is eye-catching and appealing, presenting various categories of clothing types in pictures and different panels. The webpage is also not that long, which encourages users to click on the panels present. There are also a variety of tabs that drop down into longer menus, when you hover your mouse cursor over them.

2   The news story is well presented. It includes a variety of subheadings, different fonts and font sizes, and the use of images and simple plots and graphs, which most people should be able to understand.

3   The news story is one-sided and descriptive (potentially to keep it short, with the aim of increasing the number of users who read the entire article), lacking any

sort of explanation as the why the numbers of shops on the high street are declining. For example, lockdowns due to Covid-19, and the natural decline of physical spaces for shops are not given due attention. In addition, many more retail shops are found online (like ASOS), which the article fails to mention.

Develop your skills! 7.6

1   This one is down to you!
2   The article is more of an opinion piece. There are several points where the author draws on certain historical facts and events, like when cats were domesticated, and when wild cats encountered humans, but these are not referenced to other sources, so it remains difficult to decide whether these facts are reliable and trustworthy. There are also excerpts of comments made by others on social media platforms, which work well and are woven in nicely to the authors narrative.
3   No, probably not. As mentioned above, it is an opinion piece and the author (albeit clearly biased since she is a cat lover!) does present some information on dogs. The article could be improved (i.e., by referencing facts and other data) and might benefit from including an infographic or plots and graphs to make it more engaging and appealing.

Chapter 8

Develop your skills! 8.1

This task requires you to note down your own answers and communicate what you have written to family, friends and/or pets! In terms of thinking about getting your family/friends to understand what you have said, think about the clarity of your explanations, and if they make sense to you. You might want to use examples from the real world or from shared experiences to help you in this task.

Develop your skills! 8.2

1   The research group carried out the experiment three times, probably to help strengthen the claims they made from their findings. Since they repeated the experiment and probably came up with similar results each time (since they came up with an overall conclusion from their experiments) this can help to strengthen the internal and external validity, and the reliability of their work.
2   The response variable is what the participants purchased, or perhaps also participant satisfaction. The explanatory variable is the amount of choice given to the participants in the experiment.

3   These steps were most likely undertaken to see if any results they find in the laboratory are also replicated in the real world. Working with human participants often includes many confounding factors, so any results observed that can be replicated in different environments strengthens the case that the explanatory variable (in this case the amount of choice the participants had) can explain (or 'causes') a change in the response variable (in this case how much the participants purchased).

4   This part of the task requires you to note down your own answers. You could state that the results are surprising, reflecting your own observations and experiences (links to the next question).

5   To refute the claims made, you could argue the study is limited (only three experiments were undertaken). From the abstract, however, there is no way of knowing how large the experiments are in terms of participant numbers. You could also state that if you go into your local supermarket, shop online in your favourite department store, or look at music and TV programmes to stream and download, choice is ubiquitous (i.e., it's everywhere!) and probably increasing.

   To support the claims made, you might say that the study appears to be well designed and was repeated (three experiments were undertaken).

Develop your skills! 8.3

1   Since the research aimed to look at factors that can influence media multitasking behaviour in relation to watching TV, using an observational study is practical, more cost-effective and appropriate.

2   This part of the task requires you to note down your own answers. Make sure you fully explain your argument, and points of view.

3   A problem inherent in observational studies is that many confounding variables can potentially contribute to any results reported. Other factors that could explain the results (from just reading the abstract) in this study are the presence of other people in the room or setting, the age of the participants, the gender of the participants, whether the participants re in a relationship, how many friends the participants have on social media and the frequency with which they communicate. This is just a suggested list; you may have other listed factors.

Develop your skills! 8.4

1   This part of the task requires you to note down your own answers. Make sure you fully explain your argument, and points of view.

2   Yes. Non-response bias is a possibility (since the questionnaires were self-administered, and the participants filled them in themselves). Behavioural considerations are also a potential source of non-sampling error.

Chapter 9

Develop your skills! 9.1

1   These answers will vary depending on what you need to define. You might have chosen affective, implicit, intrinsically etc. Make sure you keep a copy of all words you need defining in a small book or document and update regularly.

2   Again this will vary. However, your two sentences might look something like this: A research group carried out an investigation into several factors that can affect human choice, performing three experiments in a laboratory setting involving the purchase of gourmet jams or chocolate, and undertaking optional class essay assignments. The research group found, contrary to previous research, that people are more likely to purchase items, or undertake optional assignment work, when offered limited choices, as well as being more satisfied with their selection.

3   Previous research suggests that more choice leads to greater satisfaction, but this research group found the opposite. From my own personal experiences, I would say this is surprising!

4   Some of the terminology used might be confusing, for example 'intrinsically' or 'affective'. After you have read the abstract several times, and found definitions for the terms you were unfamiliar with, hopefully it should make more sense.

Develop your skills! 9.2

1   This abstract is a little longer than the one presented in Develop your skills! 9.1, so summarising this one could be a little trickier. There are a few ways you may have approached this task. You could have missed out the first few sentences and gone straight to the study approach, highlighting what was being investigated. The results presented are quite lengthy so you may have missed out some of these details and presented a more succinct version.

2   The journal article is well presented, and includes most parts you would expect. The introduction is called 'Background', and this is subdivided into different sections outlining how these topics are relevant to the area of study. The Method section is well presented and indicates the types of variables that were being measured and investigated. The results are presented well, supplemented with a series of graphs and tables, as well as various inferential statistical analysis undertaken. This part might be difficult to understand and require reading over several times to fully digest the contents. The final section is the discussion, which explains the results section well. This journal article does not have a separate conclusion, but what you might expect to find in a conclusion is found in the discussion in this paper.

3   The Introduction is split up into subheadings, highlighting their importance and
     relevance to the current study. The research questions for the study are then
     synthesised from the background sections, presenting three.
4   The statistical analysis might be a part that you find difficulty to understand. Try
     going over it piece by piece, and break it down into smaller components. For
     example, try to decipher the descriptive statistics first, then move onto the
     inferential areas.
5   The abstract could be written more concisely; it is a little on the long side. A
     separate conclusion section might also be useful to help summarise the main
     findings and recommendations for future research. The figures and graphs
     presented could also be improved (e.g., Figure 13.1) by using colour. Also look at
     the axes in Figure 13.1. Why are the numbers on the x-axis displayed to three
     decimal places?

Develop your skills! 9.3

1   The abstract is presented in a disjointed fashion, which is like a condensed
     version of the full paper. The authors could have summarised each of the sections
     into a single paragraph, which would be more consistent with how most abstracts
     are presented in journal articles. Most journals require the abstract to be a single
     paragraph. Some also specify a word limit of 150–200 words.
2   Here are some prompts/ideas to ponder and help you check your answers, in
     relation to this question, assuming you can access the whole article:

Why is this study important? What is the objective?

As discussed in Question 1, the abstract could be improved in terms of its layout
and purpose. The introduction is on the light side and should draw on a wider
range of references to support the development of research ideas and questions.

What was the authors' overall plan to investigate the area of interest/subject-
matter of the article?

The Methods section has more detail than the Introduction, and explains the
sample section criteria, how the sample size was determined, and the validation
of the survey instruments used. The survey used doesn't seem to have been
included; this would have been useful, especially if others were to replicate the
methods used in the study. An overview of the statistical methods used is also
included, albeit brief.

What were the authors' results and explanations for them?

Table and figures used are clear and present the data well. The confidence interval
stated could have been included in the main body of the results, and the
chi-square results could have been explained better. Since the survey is hard to

find, it makes it more difficult to interpret the results, and compare them against what the participants were asked.

What are the main points from the Discussion and Conclusion?

The discussion is written to a useful length; however, the inferential statistics are not mentioned explicitly enough. The Strengths and Limitations sections are also brief and should be expanded upon. For example, what does 'Major limitations include the observational design and the small sample size' mean? Was the sample size too small, or too big? Was the design not done well? Would an experimental approach be better? Why?

The conclusion has a mixture of causal statements, and more suggestive statements based on likelihoods. This is confusing, and when reflecting on the methods selected for this study (observational) the authors should have been more careful with the language they used in this section.

3   The causal claims made by the authors, both in the main body of text and in the abstract, are definitive and use words like 'prove', which is unscientific. Also the study involved observational approaches, whereby casual claims cannot be made, which, as discussed back in Chapter 8, should be avoided with this type of study.

Develop your skills! 9.4

1   The table is presented in a typical journal article style, where the vertical lines to create columns have been removed, and the horizontal lines for each level of measurement have also been removed. In some instances this works fine, but in this case it does make it harder to read the data horizontally. The table is labelled, with the data and sample size reported. The table itself has column headings, as well as units in brackets under the levels of measurement.
2   Since there are no horizontal lines for the levels of measurement, it does make it harder to read the data.
3   The addition of vertical lines would help improve reading the table. Also only the male number has been reported, and you must work out the female number yourself; this could be added to make it easier for the reader to follow.

Develop your skills! 9.5

1   The bar graph is clear; however, the x-axis could have been labelled (even though it is described in the figure caption). There are also error bars included, which could have been explained in more detail in the accompanying text. The values on the x-axis could be presented to two decimal places. And the highest values for each of the bars in the graph could have been reported.

2   The bar graph is not explained as well as it could be, whereby only the *P*-values
     are mentioned. The authors could have done a better job of comparing the
     differences across dayparts; only the first two are compared, with the last part of
     the day explained separately.

3   Yes, label the *x*-axis and compare the different dayparts in more detail, and with
     greater clarity (i.e. the relationship between them: the *x*-axis decreases throughout
     the day). This could have been achieved by reporting the highest values for each
     of the bars in the graph, when comparing them. The values on the *x*-axis could
     be presented to two decimal places.

Develop your skills! 9.6

1   The four points present some useful tips, as a guide to help people navigate the
     statistics they may find in published journal articles. They touch on many of the
     common areas that are likely to be reported in the results and discussion section,
     from a study.

2   The level of English presented in the section (and journal article overall) is not
     good and needs to be improved. For example, what does 'simple statistic' mean
     (in the first bullet point)? They should say 'descriptive statistics' and use the
     common names most people will be able to identify. In the third point, the
     end of the last sentence, 'it simply means no difference among groups', needs
     to be rephrased and rewritten with the correct statistical language and
     interpretation. This claim cannot be made without looking at the confidence
     interval.

3   Since examples are chosen from the discipline of biology (a blood pressure
     example is used in the third bullet point), physiology and biochemistry, it
     is likely that these disciplinary areas are the intended target of the
     authors.

Develop your skills! 9.7

1   There are several parts in the text that are a little ambiguous and confusing, with
     reference to the statistical language used. For example, what does 'significant
     main effect' mean? The interpretation of the *P*-values also needs to be improved,
     especially since they have used hypothesis testing. The hypotheses they have
     produced could also be rewritten to be clearer, and in the form of a null and
     alternative (like in Chapter 4)

2   The journal article could be improved by correctly interpreting the *P*-values and
     with correct reference to 'statistical significance' or 'practical significance' as
     appropriate.

Chapter 10

Develop your skills! 10.1

1   This answer will vary, depending on what it means to you.
2

- Predict audience interest
- Increasing audience acquisition
- Reducing audience churn
- Improving ad targeting
- Monetising content
- Developing new products

You can add more detail for each subheading if you wish. Try to paraphrase what's on the webpage if you decide to include more detail.

Develop your skills! 10.2

1   These will be your own impressions of the webpage.
2   Again, these will vary depending on what it means to you. You could mention the level of detail, especially when you click on one of the coloured tiles. You might also be surprised at the actual detail contained in the tiles (i.e., written in them, before you click on them).
3   Mostly percentages, Likert scale responses, multiple options to subdivide the data from the samples taken. Data can also be explored on a week-by-week basis, or narrowed down by Age, Working status and Gender. Lots of options to explore data, with the Excel files also available (although the way they are presented makes them problematic to use with software like iNZight – they would have to be modified to be used). There are also lots of bar graphs and side-by-side bar graphs since the data is based on attitudes (with many categorical variables). The 'Go back to key findings' tab/icon, after you click on one of the tiles on the landing page, could be tricky for some to spot, since these second pages (behind the tiles on the landing page) are very busy and contain a lot of information, and options to manipulate the data.
4   This answer will be down to you, and what you think (which will partly be founded on the types of people you interact with your experience of looking at and using/interpreting data). Most the data presented is straightforward (e.g., percentages, age ranges, variables with simple levels of measurement). When you click behind the tiles on the landing page (as mentioned in the answer in Question 3) the webpages are very busy, with many options to manipulate the data available. This could overwhelm people.

Develop your skills! 10.3

1   Checking for accuracy of facts being presented is a good place to start, which can be undertaken by doing Google searches, using reputable sources of information. For example, official government websites are usually trustworthy, as well as sites such as the BBC, NHS, and other organisations like the WHO.

Questioning the accuracy of an article may arise in relation to where you find it. For example, certain social media platforms (Instagram, Twitter, or Facebook) are likely to include accounts of personal opinions, with certain agendas. These may be biased, depending on their motivations, etc. Motivations might be political, or the article might be trying to sell you something. It's useful to always question the motivation behind an account or social media post, to give you a better idea of where the author is coming from and why they are trying to convince you of a certain perspective or point of view.

2   Fake news exists mainly to convince you of the author's perspective or standpoint. If there is a lack of evidence to support their claim, then fake news is a convenient avenue for them to exploit. The link for this question takes you to a webpage that has changing stories, which will influence what you write for this question, but it is a useful page populated with many examples of fake news. This page represents a worrying trend, and the reality that we live in a world with increasing amounts of fake news. This means we need to be increasingly vigilant and critical of accounts and data we see, and we need to question its accuracy, validity and reliability (some of these concepts are from Chapter 8).

3   Sometimes fake news stories come from reputable websites or social media platforms we think we can trust (e.g., Facebook), which brings into question their role in policing who can say what on these platforms, and the tensions that can exist with freedom of speech laws. Certain outlets and authors can gain support and traction depending on external/societal pressures. For example, after the recession of 2008, and ongoing issues with rising unemployment in the USA, more far right and extremist views were deemed to be much more accepted by many Americans in swing states (i.e., states that can swing either way to Republican or Democratic majority support). On top of this, Donald Trump's victory in becoming president in 2017 can arguably be accounted for due to his team's use of fake news, to discredit the Democratic candidate, Hillary Clinton. Donald Trump is seen as one of the modern architects of fake news (and has had his Twitter account suspended on several occasions for posting factually incorrect accounts and tweets), and cleverly shut reporters and other media outlets down by claiming they were fake news.

Motivations for fake news from apparently reputable people can be due to attention seeking, or trying to convince people of their views, which can be extreme or unjustified and unsupported by the relevant facts available at the time. Another example of this is in relation to the Covid-19 vaccine, which attracted a lot of fake news stories and attention. A small minority of nurses and doctors have recommended (in YouTube clips and social media posts) that people do not take the vaccine, which has been scientifically proven to save lives and protect individuals from infection and serious illness. These kinds of reports can be especially damaging, since you might ask yourself why these people would want to mislead and potentially harm others. Sometimes it is not clear why certain people or groups deliberately try to mislead others; the best we can do is look to trusted sources of information (like the NHS in the case of Covid-19 vaccinations) and question the existence of fake news stories, after we have looked at the relevant stories and reports with a critical eye, checking on the accuracy, validity and reliability of the information and data presented.

Develop your skills! 10.4

Here is a suggested revised version that improves the data story's clarity and narrative (yours might be different, this is just a guide as to how an improved version could look):

IS BREXIT STILL OF PUBLIC CONCERN?

Public opinion on the vote for the UK to leave the European Union (EU) has fluctuated since the official vote on 23 June 2016 (NatCen Social Research, 2019). It appears that there is now a greater proportion of remain voters (65% from a sample of 3563) in a recent poll taken on 21 July 2019, with the leave camp now in the minority.

Many people feel like they were lied to by the leave campaign, before the official vote took place. But the question now remains: do people still care about this important issue? Has it become too political?

A recent poll conducted by Wake-Up UK Live (WUUKL) asked a sample of 321 people on the streets of London, 'Do you still care about BREXIT?' The poll results revealed that approximately 72% said No, with 25% answering Yes and 3% answering with a Don't know.

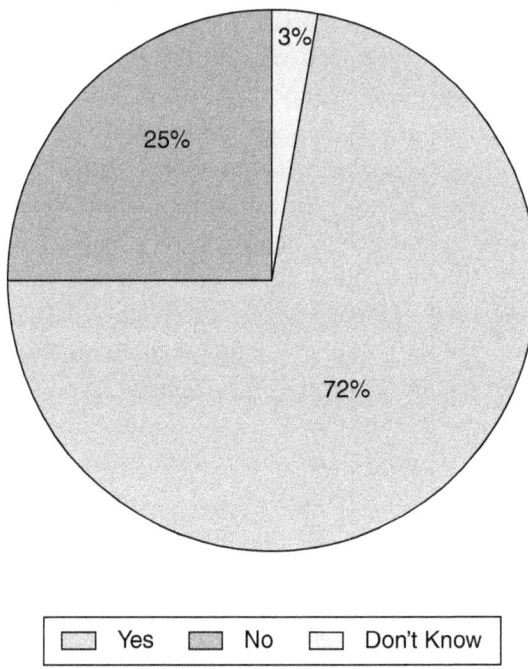

So, can we conclude that this issue is still of public concern? More data is needed from people across the UK to probe this further.

NatCen Social Research (2019, 18 November). What UK thinks EU: Non-partisan information on UK attitudes to the EU before and since the EU Referendum. https://web. archive.org/web/20191216023219/https://whatukthinks.org/eu/questions/if-there-was-a-referendum-on-britains-membership-of-the-eu-how-would-you-vote-2/

Develop your skills! 10.5

1   The article's title suggests a bias towards the high street being in crisis, but there are some counter-points made in the article. For example, the article includes numbers of retailers opening and closing from 2016 to 2021. It also includes a breakdown of the types of retail sectors growing and declining from January to July 2021. There is also a side-by-side bar graph, titled, 'The high street crisis continues'. The article also presents a specific case study of a retailer which has been successful during the pandemic, and there is also a section called 'Green Shoots' which suggests there is still hope for the high street, and that shoppers want an in-person shopping experience. So, at first glance it seems like a worst-case scenario, a bit of doom and gloom, but overall, the article is mostly balanced.

2   Building on your answers in Question 1, there are three graphs/plots that are presented in the article. The first two plots are understandable and presented well;

however, it would have been useful for them to fully name the acronym PwC, which supplements these plots. The final plot is less clear and doesn't give an indication of when the data were collected, or over which period they were referring to.

Chapter 11

Develop your skills! 11.1

1   The article tries to make the case for the inclusion of an alternative to the theory of Intelligent Design, to be taught in schools.
2   It appears to be a comical article, since the claims made and the examples used (Flying Spaghetti Monster, pirates and Pastafarianism) seem too extreme and unusual to be serious. The inclusion of the artistic drawing at the end of the article also adds to the humorous nature of the webpage.
3   The article appears to be making similar claims to people who believe in Intelligent Design, to help identify errors in their logic and reasoning.
4   The evidence presented, and data included, seem to purposely emphasise the ludicrous nature of the claims being presented. The graph showing a correlation between numbers of pirates and global average temperatures suggests a causal relationship, which seems to be purposely unbelievable. The relationship between the variables in the article is presented as being a statistically significant inverse relationship, which cannot be ascertained from the graph presented. This is also an incorrect description of linear relationship since no t-test has been performed. Data and numbers are presented using statistical terms like 'approximate' and 'statistically significant'. Sadly, these are presented in a vague or incorrect way (approximate numbers of pirates, for example). There are also no references or links to other work linked to the arguments being made, which would help to improve the legitimacy of the claims being made.

Develop your skills! 11.2

1   Enjoy reading the article you chose!
2   This answer will depend on the news story/article you chose to read. Most of the stories are well presented, and are broken up with subheadings, and the use of different fonts and font sizes to keep your attention. They are also short, so you do not have to worry about spending too much time reading them.
3   Again, this will depend on what you click on to read. Most the articles summarise big news stories that are likely to have fake news aspects, or disinformation/misinformation circulating on social media platforms and online. There are also links to fake news stories which the BBC News website tries to investigate and provide the reader with the truth.

4   The news stories are based on the BBC News website, which is a well-respected, long-established and trustworthy organisation. There is also embedded video footage of certain events, linked to the news stories, used to refute fake news.

5   Again, this will depend on what you click on to read. Clues might include a survey that was created and distributed. There will likely be interview data presented, with quotes included. There may also be data sources included, indicating the agency or organisation that collected and potentially analysed the data, to help attribute credibility to the claims being made in the news story.

Develop your skills! 11.3

1   There is a mixture of quantitative and qualitative forms of data presented in the article. The qualitative comments included towards the beginning of the article are inserted to exemplify related points being made. Other data and information are included using references to news stories, YouTube clips and supporting evidence presented by academics from a range of universities. Twitter excerpts are also included. These forms of evidence are mostly qualitative and are eye-catching. Results from experiments or research findings, that have been peer-reviewed, are not included, which would have helped improve the validity and reliability of the data presented.

2   Since the article is lacking evidence from reliable and valid sources, the arguments made come across as biased and on the weak side. The various comments and qualitative evidence presented, along with the images and YouTube clips included, are engaging but lack substance.

3   Overall, the language used to present the data included in the article is good. Although there is a lack of data and statistical information included, the use of words like 'approximately' and 'about' sets an appropriate tone.

Develop your skills! 11.4

1   Yes! In terms of statistical literacy and helping the reader to make connections with the data being presented, option 2 is better.

2   Yes, there is no abstract. This should be included to help people search for articles, to give them a synopsis or overview of the content in the article.

3   There are several mistakes made in the bullet points (e.g., in the first point, the last sentence should start with 'Presenting', not 'Present'). The points are also a little vague. For example, what does 'keep it simple' mean? This will really depend on the context and the data.

Develop your skills! 11.5

1   There are several issues with the plot. The x-axis is reversed (larger values moving across the smaller values), which is unusual. The x-axis has uneven increments

moving across, which is not good practice. The *x*-axis is displayed as number of pirates (approximate), which is ambiguous. In addition, there is no indication as to where these pirates were counted – are they global figures? Along the line on the plot, the year date is overlaid, and this alternates between 20- and 40-year increments.

2   Global temperatures could have been ascertained using data from the relevant meteorological webpages and other sources. The number of pirates seems a little harder to locate, with pirate numbers appearing in several newspaper articles (after a quick Google search). Nothing concrete here, however, so it is a little puzzling as to how these data were collected.

3   The two variables are inversely correlated.

4   No, there does not seem to be a causal relationship between these two variables. It is hard to imagine how a decrease in pirate numbers could lead to an increase in global temperatures.

5   Using the plot associated on the webpage, after clicking the link, it is unclear how the author can make such a claim. Further statistical analysis, including inferential tests, would be needed to make the claim stated in the article.

Develop your skills! 11.6

1   The main reason being presented seems to point to the complexity of what a *P*-value measures. And the way it is often interpreted in journal articles and other outlets is either too simplistic or incorrect.

2   Improved training about what a *P*-value can and can't do would help to improve the communication of statistical significance. This could be achieved by threading key statistical concepts in the early years of schooling and carried on throughout school.

3   The conclusion cautions against overinterpretation of observed associations, with reference to the *P*-value.

Chapter 12

Develop your skills! 12.1

1   Yes, the section describes the research group and associated study that contains the data used in the research study (Video Consumer Mapping study). However, the link takes you to the incorrect journal article (it's linked to the same journal article being reviewed in this question). A list of 495 adults were recruited from six areas. This section is not written well, and it's not entirely clear how the CRE and VCM are linked. The method used to recruit participants is also not described, were they randomly selected?

2   Genres = categorical (nominal), Day parts = categorical (nominal) and Social viewing = categorical (nominal).

3   Multimedia tasking = Discrete.

4   No.

5   Since the assumptions to use ANCOVA are not met, it is inappropriate to use this
    test. The language used to describe the results from the statistical analysis is also
    poorly communicated. This part is especially poorly worded:

    Overall, the model was significant. Significant effects are reported at the 99%
    confidence interval (i.e., $p < 0.01$) because of the large number of observations.

    What does the model was significant mean? Since the assumptions to use
    ANCVOA have not been met, this is not an appropriate model to use. Significant
    effects is also an erroneous way to report a P-value.

Develop your skills! 12.2

1   The webpage is quite long and requires a lot of scrolling. It is also littered with
    distracting adverts. The introductory paragraphs give a clear indication as to
    the main points. Certain parts of the introductory information are not clear,
    for example what does this mean: 'Employers' reduced degree requirements for
    middle-skill positions and 31 precent of high-skill positions were reduced by
    46 percent between 2017 and 2019, the report said.' Having read this
    paragraph ten times, I am still struggling to ascertain what it is trying to
    convey!

    The data presented for each job type mostly include four bolded variables, but
    some have only three. Upon closer inspection, this looks like a potential error on
    the webpage. For 'real estate agent', after the third variable presented, the fourth
    one (Number of people employed in the US) is in non-bold type. It should be in
    bold and separate, for consistency.

    There is also no indication as to where the data have come from to support the
    claims made and information presented, in the article.

2   As mentioned in the answer to Question 1, the data presented for each job type
    mostly includes four bolded variables, whereas some only have three. There
    seems to be a lot of inconsistency when looking at the values reported for each
    job type. For example, the median annual wage for flight attendants is reported
    as '$59,050, with the top 90 percent of flight attendants making over $115,000
    per year'. What does 'the top 90 percent of flight attendants' mean? And how
    much more than $115,000 per year are they making? This is not clear.

    Then for the next job type, 'theatrical and performance makeup artist', the
    median annual wage is given as one value. Why is there not a range given?

3   Building on the answer for Question 2, they are not explained well, and there is a
    lot of inconsistency between job types, for all the variables stated. The four
    variables used are:

    i   Median annual wage: Interval
    ii Level of difficulty: Ordinal

iii   Location and/or remote: Nominal (binary in places)

iv   Number of people employed in the US: Discrete

As mentioned in previous answers, there are inconsistencies (e.g., some wages are reported as a specific value, while others are given as a specific value and a range. No explanation is given as to why this is the case).

4   The statistical language used is poor and demonstrates a lack of basic statistical literacy. In several places, the word 'over' is used in conjunction with data values (e.g., 'Number of People Employed in the U.S.: Over 248,700'). Part way through the webpage, numbers are qualified with the word 'approximately' (e.g., for bartender: 'Approximately 485,330'). This seems like a very specific value for it to be an approximation. Then for the last job listed (personal trainer) the following comment is made: 'Median Annual Wage: $46,245 to $78,747, with the average salary of $64,100, but statistics vary significantly by location as larger cities offer significantly higher salaries for personal trainers'. A median is one number, not a range of values.

5   No tables or graphs are included. Since there is a lot of data presented, the addition of graphs would make the article more eye-catching and engaging. A bar graph would have been useful to help compare median annual wages of the jobs listed. A table could have been used for jobs listed and number of people employed in the US.

6   No, since there are so many mistakes littered throughout the article. In addition, the language used is poor, and there are no graphs or tables, which makes the article dull and lacklustre. The title of the article is also misleading, since it suggests the highest-paid jobs don't require you to have a college degree. What the article is showing you are the highest-paid jobs for people who don't have a college degree, but there are many jobs that pay higher salaries compared to the jobs presented on the list, which require you to have a degree. The data source is also not clear, which casts doubt on the credibility of the data presented.

Develop your skills! 12.3

1   The y-axis starts at 94 million and finishes at 100 million. The differences between the quarters in each year and for the entire x-axis are therefore misleading, since the differences are made to look bigger. It looks like the data for 2009 Q1 is half the value of 2009 Q3.

2   They wanted to overinflate or sensationalise the number of people receiving welfare benefit in the USA in 2009–11. Since this is from a news article outlet (Welfare), it's likely that the author wants to engage and attract readers to click on this story and read it.

3   Yes, this is a pie chart, and the segments don't add up to 100%.
4   A bar graph would have been a better presentation of the data displayed in
     Figure 12.5.

Develop your skills! 12.4

1   Yes, there are several parts that are unclear. For teachers (and general readers!)
     unfamiliar with significance, what does 'significantly different' mean in the first
     sentence? The advice given under the graph that displays salt content is also not
     described well. After stating that the outlier has been confirmed to be a legitimate
     data point (i.e., not an outlier), advice is then given to remove it and treat it as if
     it were an outlier. This seems to be counterintuitive.
2   It depends on the context, that is, what the data is showing you. The formula
     presented will not tell you (in relation to the context) if a data point is an outlier, it's
     more of a rule of thumb. This is also mentioned on the webpage used in Question 1.

Develop your skills! 12.5

1   Overall the paragraph is clear and presents the descriptive data directly. What is
     not clear are the results for males and females. Are the stroke risk factors for
     everyone or males or females?
2   It should be made clearer whom the stroke risk factors are referring to.
3   The same issue exists in the table, as with the introductory parts of the Results
     section. The table description mentions that it refers to the Aeolian population,
     so it is not clear if this is for everyone or males or females. In addition, the values
     for Age and the % column aren't clear. In the introductory part of the results, it
     mentions the mean is 52, in the table this is under the % section. The table also
     mentions Sex (male) with the percentage and then lists the stoke risk factors. So
     where are the values for females?
4   The issues in the answer to Question 3 need to be addressed. The table should
     include values for males and females or make it clearer that these are for all
     participants. The % column heading should also be made clearer.

Develop your skills! 12.6

1   The authors have tried to be consistent by including a decimal point for whole
     numbers; however, this is not necessary. One data point is rounded to 60% (a
     scanty use of cured meat) and not presented as 60.0%, and in the table (Table
     12.2) it is presented as 59.9%. The presentation of data also leaves the reader to
     cross-check with the tables and do a lot of work (i.e., it just says data are shown
     in Tables 3–5 then describes the data).

2   It's not clear why one data point, a scanty use of cured meat, is presented as 60%, when in the table it's 59.9%.
3   Apart from the issues mentioned in the answers to Questions 1 and 2, there are broken lines for the variables Eggs and Sweets, in Table 2. This should be one line, to be consistent with the variables in the table.

Develop your skills! 12.7

This task will depend very much on what you select to investigate, and the data you collect as a result. I hope you very much enjoyed this task and managed to investigate something that you are interested in, to showcase all the skills you have picked up in this book!

Chapter 13

Develop your skills! 13.1

Both questions in this in this task will depend on the examples you chose. Your answers will involve personal reflection on the skills you have begun to develop in this book.

Develop your skills! 13.2

All questions in this in this task will depend on the examples you chose. Hopefully you found this exercise useful!

Develop your skills! 13.3

This task will depend very much on what you choose to investigate and the data you collect. I hope you very much enjoyed this task and managed to investigate something that you are interested in, to showcase all the skills you have picked up in this book!

Develop your skills! 13.4

1   Webpage (a) seems to be targeted at people involved in the world of business. Several of the claims made in the webpage have links to other pages, but these don't seem to be backed up by any reliable research. There are also links to Venngage, which is the platform being presented in webpage (c).

    Webpage (b) is a link webpage (a). The writing is persuasive but falls down due to a lack of links or references to reliable and valid research to support the claims being made.

    Webpage (c) is linked to a company that seems to make infographics for users. It is well laid out and includes useful information, as well as embedded video clips to support the written content, and other data displays. There are also links to data sets and other data displays that seem legitimate.

2   In webpage (a) there are clear biases, with the following claim being made about the author: he is dedicated to helping internet retailers succeed online by developing digital marketing strategies and optimised shopping experiences that drive conversions and improve business performance. This reads to me like there might be costs involved for the services offered on the webpage! However, it clearly spells out the target audience: internet retailers.

Webpage (b) is provided by Directive Group, which offers digital marketing solutions for internet retailers. The discourse used in this blog, and the writing style utilised, are both aimed at internet retailers, hoping to persuade them of their arguments, which seem to lead to the products they have to offer. Its ultimate aim is to advertise its services.

Webpage (c) offers free templates for others to use to help create their data stories. Since it is linked to Venngage, it is clearly looking for people to pay for some of their services. The top left part of the webpage does state that sign-up is free, which does mean that users who do sign up can pay for further support and resources.

3   Webpage (a) offers useful advice and guidance, webpage (b) includes some interesting points, and webpage (c) includes a lot of useful information, but all would be more persuasive and believable if they were supported by journal articles or other research, to support some of the claims made.

GLOSSARY

Alternative hypothesis   The alternative hypothesis is sometimes called the research hypothesis. It is usually written to explain that something is happening, there is a difference or an effect, there is a relationship.

Association   If the data points on the y-axis increase in line with the x-axis, then we can say there is a positive association. If, however, the data points decrease on the y-axis as the x-axis increases, then we say there is a negative association.

Bar graph   A bar chart (or graph) organises information into a graphic using bars of different lengths. The length of these bars is proportional to the size of the information they represent. For example, a vertical bar graph can show the popularity of different colours among a group of children, or the main car of choice of countries in the European Union. Bar graphs are extremely useful for displaying data from a table of counts.

Behavioural considerations   People tend to answer questions in a way they consider socially desirable. For example, pregnant women being asked about their drinking habits are unlikely to admit to consuming alcohol, and people are more likely to say they have never cheated on a partner, even when they have.

Bias   Bias in statistics refers to a systematic error in the collection, analysis, interpretation, or presentation of data that leads to results or conclusions that are systematically different from the true values or population parameters. Bias can occur at various stages of the statistical process and can significantly impact the validity and reliability of study results.

Biased selection   A biased selection process is an inadequate selection or sampling process, examples of this can include selection bias and non-response bias.

Bootstrap confidence interval   A bootstrap confidence interval for a parameter can be thought of as a range of plausible values for the true parameter value.

Box plot   The box plot is a useful form of display since it shows the centre and spread of data. It usually reveals the median, interquartile range, and lower and upper quartile values.

Categorical   Categorical variables are often defined by words. They can be ordinal or nominal. Ordinal variables have a natural order or hierarchy – for example, level of education (BSc, MSc, PhD) and income (low, middle, high). Nominal variables have no order   for example, gender, type of car, occupation.

Causation and correlation   When looking at correlated data (Chapter 4), we can often say that there is an association between variables, for example as one variable increases so does another, and if the data points related to these variables can be modelled using a straight line, we may call it a linear relationship. This relationship might lead us to assume that a change to one thing *causes* a change in the other. Reasons for this include a human need to simplify incoming information, so we can make sense of it. Our brains often do that by making assumptions about things based on slight relationships, or bias. But that thinking process is not foolproof.

Cluster sampling   This entails dividing the population into clusters, such as geographic regions or schools, and randomly selecting entire clusters to include in the sample. This method is useful when it is impractical or costly to sample individuals directly.

Confounding variables   Experiments rely on the researcher controlling certain variables and then manipulating others, to see if there are any causal relationships present. Controlling certain variables can be difficult if not impossible – these are what we call confounding variables.

Convenience sampling   This involves selecting individuals who are readily available and willing to participate, making it a convenient but potentially biased sampling method.

Data storytelling   This is a structured approach to communicating data insights, and it involves a combination of three key elements: data, visuals, and narrative.

Descriptive statistics   These are used to describe the basic features of a data set. They provide simple summaries about the sample or population. With descriptive statistics, you are simply describing what the data show. Descriptive statistics can also be broken down into measures of central tendency and measures of variability (spread).

Disinformation   Disinformation is false information that is deliberately created and spread to influence public opinion or obscure the truth.

Dot plot   A dot plot, also known as a strip plot or dot chart, is a simple form of data visualisation that consists of data points plotted as dots on a graph with an x- and y-axis. These types of charts are used to graphically depict certain data trends or groupings. They can be expanded to include side-by-side plots, for example when you are looking at a ratio variable (e.g., weight) subdivided by a categorical viable (e.g., gender).

Estimate   A *known* quantity calculated from (sample) data to estimate an *unknown* parameter; for example, a sample mean can be used to estimate the unknown population mean.

Ethical considerations   These are important to bear in mind when doing certain kinds of research. Most research (including observational studies) requires the researchers to apply for ethical clearance from an ethics board. This ensures that the subjects involved in the study (which can include animals, for example) are protected against any potential harm (which could be physical or emotional), and the appropriate safe-guards are put in place in case any harm does occur.

Experiments   These involve a researcher changing or manipulating certain conditions, to see what effect it has on a response variable. In an experiment, the researcher deter-mines the treatment and control groups. The inclusion of a control group, who do not receive the treatment or intervention, can help in making useful comparisons between groups. The main aim of an experiment is to establish a causal relationship, for example, whether a change in one variable leads to and *causes* a change in the response variable. In an experiment, there is also usually a random allocation of treatments to groups, to try to mitigate any bias in the data. Some studies, for example, include *blinding*, where the researchers or participants do not know which group received the treatment. This can help prevent any biases occurring when the researchers interpret the results.

Experiment-to-causation inference   When looking at experimental data involv-ing random allocation of a treatment of some kind to groups in an experimental set-up, we may want to make a causal inference (e.g., when we change one variable, say temperature, it *causes* a change in another variable, which could be rate of sweating in humans), which we call experiment-to-causation inference.

Explanatory variable   Sometimes called the *independent variable*, in the context of an experiment, this is the variable that is said to explain or predict any differences or changes we see in the response (or dependent) variable. The explanatory variable is the one an exper-imenter changes or modifies to see what effect it has on the response or dependent variable.

Fake news   Fake news is a term that we see increasingly on the news and on social media platforms; however, it can mean different things to different people. In its broad-est sense, fake news can refer to news stories or social media posts that are false: the story itself is fabricated, with no verifiable facts, sources, or quotes. Sometimes these stories may be propaganda that are intentionally designed to mislead the reader or may be designed as 'clickbait' written for economic incentives (the writer profits on the number of people who click on the story).

Fixed-level statistical significance   Fixed-level statistical significance (i.e., when researchers or news articles look at P-values) involves a cut-off point, which is usually 0.05 (5%).

Histograms   Histograms are an extremely useful and common way of representing data. They are like bar charts but show the frequency density instead of the frequency. They can also be used to determine information about the distribution of data.

Hypothesis   A hypothesis is a tentative statement that proposes a possible explanation to some phenomenon or event. They are often used in deductive reasoning, as part of a scientific methods approach. Note hypotheses in this chapter are posed as statements, which can be useful to develop a hunch or idea, often based on existing knowledge and/or data. They can be further developed and converted into null and alternative hypotheses, as presented in Chapter 4.

Hypothesis testing Hypothesis testing is a statistical method used to make inferences about population parameters based on a sample of data. It involves the formulation of two competing hypotheses about the population parameter of interest, and then using sample data to determine which hypothesis is more supported by the evidence.

Interviewer effects Different interviewers asking the same question can obtain different results. This could be due to the sex, race, or religion of the interviewer. For example, when asking male participants 'How often do you feel depressed?', a female interviewer found that 21% answered 'Often', while a male interviewer found 6% gave that answer.

Knowledge Knowledge exists in many forms and can be viewed as the basis of humanity's desire to develop expertise in a variety of subjects. The approaches taken to generate and analyse the data needed to reaffirm and create new knowledge are partly determined by the discipline in which it sits. For example, chemists and biologists often follow the scientific method to generate data to support or refute a hypothesis they have come up with to solve a problem or fill a gap in knowledge. They often choose experimental methods, whereby they control certain conditions and manipulate others to generate quantitative data. In using this approach, these scientists can replicate experiments to help improve the reliability of their results. In addition, if they can see a similar patterns over time, scientists may be able to generalise these results externally (called external validity).

Levels of measurement Levels of measurement, also known as scales of measurement or measurement scales, refer to the different ways in which variables can be classified and measured. Examples of this include: Nominal, Ordinal, Categorical and Numeric. Researchers need to ensure that they are aware of the levels of measurement they are working with, to ensure appropriate statistical techniques are used.

Mean This is a statistical name for the ordinary, everyday average. It is where the dot plot balances. It can be calculated by adding up all the values of your data points and dividing by the number of data points you have.

Median This is the middle value in your data.

Misinformation Misinformation is false or inaccurate information that is mistakenly or inadvertently created or spread; the intent is not to deceive.

Modality (or mode) The modality of data refers to the most frequent points in your data set. If there were two clear modal values, and the data spread looked like a camel's back with two humps, we would say the data has a bimodal distribution.

Non-response bias This is due to choosing a certain group of people to be surveyed but who do not respond. For example, they may refuse to participate or be unreachable.

Non-sampling errors These can be much larger than sampling errors and are always present. They are often impossible to correct for after the poll or survey has taken place. Any potential non-sampling errors must be minimised in the design phase of the poll or survey.

Normality-based (or norm-based) confidence interval This is a method for generating an interval for a parameter, which involves a range of plausible values for the true parameter value. In this case the data are assumed to be normally distributed.

Null hypothesis The null hypothesis usually takes a doubtful point of view: the researcher's hunch is unconvincing, there is nothing new or interesting happening, there is no effect. In other words, the results from say a well-designed experiment are just due to chance, and have nothing to do with any factors, i.e., any changes we see in one variable are not being caused by the researchers changing another variable.

Numeric Numeric data can be discrete or continuous. Discrete data are often counts – for example, number of pies sold, number of caps worn in a month. Continuous data are often measurements and can be on either a ratio (e.g. weight of rabbits, height of trees, heart rate, distance, age) or interval scale (e.g. temperature in degrees Celsius, pH).

Observational studies These often involve a researcher observing groups of people, or animals, or objects of interest, at a single point in time or over a longer period. These studies can include the collection of data using qualitative (e.g., group interviews) or quantitative approaches (e.g., surveys), or mixed methods (which could be a combination of qualitative and quantitative methods). The researcher might be involved with measuring changes to a response variable over time, looking for potential causal relationships, or they might be interested in collecting data over time to look for common themes. Observational studies can be cross-sectional or longitudinal. Cross-sectional studies provide a snapshot of data at a point in time, from a research subject of interest, whereas longitudinal studies follow the research subjects over time, which can include data collection at multiple time points. Cohort studies, such as the Caerphilly Cohort Study in Wales and the Dunedin Study in New Zealand, are examples of longitudinal studies.

One-way table of counts When a table presents data for one, and only one, categorical variable, it is called a one-way table. A one-way table is the tabular equivalent of a bar chart. Like a bar chart, a one-way table displays categorical data in the form of frequency counts and/or relative frequencies.

Outliers An outlier is a data point that appears to be far away from the rest of your data points. There is sometimes the temptation to remove outliers from a data set, which should be avoided! It is much better to ask questions about that data point. Why is it so far away from the others? Could there have been a mistake in reading the data? On the other hand, is the outlier of interest? Removing a perceived outlier can have big implications for any statistical analysis you perform on the data set and might not give you the true answers.

P-value The *P*-value gives us an idea of how likely (or unlikely) it is to get the results we did just by chance (when the null hypothesis is true). The *P*-value measures the strength of evidence against the null hypothesis, H_0. The smaller the P-value, the stronger the evidence against H_0.

Parameter A numerical characteristic of a population or distribution, such as a population mean.

Percentages These enable the presentation of data on a scale between 0 and 100%. It is always useful to know what sample or population size a percentage is based on.

Polls and surveys These are created for many different reasons and come in many forms. We are increasingly exposed to different ways of data being extracted from us, especially in digital form. For example, data collected from responses to a survey can be used to tailor emails that present you with specific offers from a retail store. Polls can be used to help predict the outcome of upcoming events, such as political elections. Polls tend to be based on one specific question, whereas surveys are larger, often including multiple questions and question types.

Population This includes all the data points of interest, which can come from a group of participants, people, plants, animals, etc. Often a subset of the population (a sample) is chosen to represent that population.

Practical significance Practical significance (practical importance) relates to the size of an effect. The size of an effect is estimated with a confidence interval (which can be generated by using either a bootstrap or norm-based method). Whether or not an effect/difference is of practical importance is a consequence of how big it is. Look at a confidence interval when determining the practical significance of an effect. The size of a statistically significant effect can be so small as to have no practical importance at all, that is to say, statistical significance does not imply practical significance.

Proportions These can be presented as fractions or with a decimal point. For example, you could say ¾ or 0.75 (simplified from 15/20) voted for the first candidate while ¼ or 0.25 (simplified from 5/20) voted for the second.

Question effects Variations in wording can influence responses. Compare 'Do you think there is a lack of discipline in children? Do you believe in national conscription?' and 'Do you think children should be given the time and space to grow, and play with other children? Do you believe in national conscription?'

Random sampling This helps to avoid subjectivity in choosing participants and allows for the calculation of sampling error. The larger the sample taken, the better, in terms of it being representative of the population of interest. Also known as probability sampling, it involves selecting individuals from a population in a completely unbiased manner, giving everyone an equal chance of being chosen.

Randomised control trial (RCT)   A commonly used experimental approach. RCTs involve the random allocation of people (or units) to receive one of several interventions. These can be clinical, sociologically based, or psychological interventions. One of these interventions is the standard of comparison or control. The control may involve a placebo ('sugar pill') or no intervention at all. A placebo should not have any causal effect on the dependant (or response) variable. Someone who takes part in an RCT is called a participant or subject. RCTs seek to measure and compare the outcomes after the participants receive the interventions.

Ratios   These describe a relationship between two numbers, indicating how many times the first number contains the second. For example, if 20 people were asked to vote for one of two candidates, with 15 voting for the first candidate and 5 voting for the second, as a ratio, this would be written as 3:1. Ratios and proportions are interchangeable, as are percentages and proportions.

Regression to the mean   This is a statistical phenomenon that can make natural variation in repeated data (i.e., obtained by taking multiple measurements of the response variable, for example investigating the effect of chocolate on heart rate, where multiple measurements of heart rate are taken) look like real change. It happens when unusually large or small measurements, which can include baseline observations, are followed by measurements that are closer to the mean.

Reliability   This explains the degree to which a research instrument (e.g., a stopwatch measuring time) measures a given variable consistently every time it is used under the same conditions with the same subjects.

Response variable   Also known as a dependent variable or outcome variable, is a key element in statistical analysis and experimental research. It represents the outcome or result that is being studied or measured in an experiment or observational study. In other words, the response variable is the variable that researchers are interested in understanding, explaining, or predicting.

Sample   A sample is a set of data collected from a population.

Sample-to-population inference   Often when we want to make an inferential statement using a sample that has come from a larger population, the sample needs to be randomly selected to help mitigate against any biases. When we take sample data like this and want to say something about the population it has come from, we call this kind of inference sample-to-population inference.

Sampling   Presenting a survey or poll to a sample from a population of interest is often faster, cheaper and more practical than trying to get data from the whole population. Poll and survey reports should include the target population, sampling method, sample size, date and the exact questions asked.

Sampling errors These occur because of taking a sample, affecting how representative it is of the population it came from. They have the potential to be bigger in smaller sample sizes.

Scatter This gets you to ask the question whether the data points are constant on a scatter plot. If you were to draw a straight line though the data points on your graph, are they close to the line? Alternatively, is there a lot of variation in data points around the line? Answering these questions explains the scatter of the data points in the graph.

Scatter plot These figures are extremely useful, in terms of the displays selected, since they can give you useful information on whether the variables in the graph are associated with each other, and they also retain the numerical information about the variables.

Selection bias This is the situation where the population sampled is not exactly the population of interest. It might occur, for example, if you asked readers of *Vogue* magazine (USA) in an online survey for their opinions on same-sex marriage via an online survey, and then tried to generalise the results to all US people.

Self-selection bias This can occur when people choose to volunteer in a poll or survey, that is, they are not randomly selected. For example, a TV show could present a poll question and ask viewers to phone in or to respond online. Since the viewers decide themselves to participate (i.e. they are not selected), this can give rise to self-selection bias.

Sensitive data This is defined as any data set that is linked to a controversial or potentially sensitive topic, that can be emotionally triggering, upsetting or cause distress.

Side-by-side plot Side-by-side plots are an extremely useful way to split a ratio variable by a categorical variable. They can help you to further explore any patterns you observe in the data. For example, a ratio variable like Weight might be bimodal, but subdividing it by a categorical variable like Gender may help explain the bimodal split (since, on average, boys tend to be heavier than girls).

Significant figures Significant figures are the number of digits in a value, often a measurement, that contribute to the degree of accuracy of the value. We start counting significant figures at the first non-zero digit.

Skewness When looking at dot plots or histograms, how the data are shaped will tell you if they are positively skewed, negatively skewed or symmetrical. If the tail end of the data is located towards the positive x-axis values, then we call this positive skew. If the tail end of the data points is located towards the negative x-axis values, then we call it negative skew.

Standard deviation This can be thought of as the typical or average distance between the individual data points and the mean.

Statistical inference The process of using sample data to make useful statements about a population, for example when using sample data to estimate an unknown population mean. You could also be looking at a sample median, and you could also be dealing with differences between sample means, a single proportion from a sample, or differences between proportions from two different samples.

Statistical literacy Statistical literacy, sometimes called data literacy, can be defined as the ability to understand and reason with statistics and data. The literacy elements require the use of good writing skills, being able to craft a coherent and concise narrative, intertwined with data and numbers. Statistical literacy often includes a range of thinking and practical skills that include knowledge, comprehension, application, analysis, synthesis and evaluation. It enables a feel for data, including being able to support an argument with evidence, but also being aware of the variety of interpretations are possible from those data.

Statistical model A statistical model can provide useful information that aids researchers in identifying relationships between variables. They can also be applied to raw data (which can be sample or population-level data) to aid us in making predictions.

Statistical significance The most common method for assessing statistical significance involves comparing the observed data to a null hypothesis, which states that there is no real effect or difference. Statistical tests generate a P-value, which represents the probability of obtaining results as extreme as the observed data, assuming the null hypothesis is true. A small P-value (typically below a predetermined threshold, often 0.05) indicates that the observed results are unlikely to have occurred by chance, leading to the rejection of the null hypothesis in favor of the alternative hypothesis.

It's important to note that while statistical significance is a valuable tool in research, it does not guarantee practical significance. Both statistical and practical significance should be considered, when interpreting their findings.

Stratified sampling This technique involves dividing the population into distinct subgroups, or strata, and then selects individuals from each stratum in proportion to their representation in the population. This method ensures that each subgroup is adequately represented in the sample, making it useful when studying heterogeneous (i.e. mixed) populations.

Strength of relationship If the data points on a scatter plot are close to a straight line drawn in such a way as to fit the data points as closely as possible, then we can say there is a strong relationship between the variables. If the data points are spread out and far away from each other, with no discernible pattern, then we say there is a weak relationship.

Survey-format effects This can occur when the layout or order of questions is presented differently to different people or groups. Examples are question order, survey layout, and whether interviewed by phone, in person or by mail.

Symmetry   Assessing the symmetry of your data points, which is best performed on dot plots or histograms, can be thought of as placing a line somewhere in the middle of the data points, and seeing if it mirrors each side. For example, if you could place a line in the middle of your data points on a piece of paper, and you folded it along the line you have drawn, would they roughly fit over each other?

Systematic sampling   This involves selecting individuals at regular intervals from a population list, which can be efficient and less time-consuming.

Tables   These are used to present data, which can highlight interesting patterns or relationships.

Test statistic   A test statistic is a summary statistic that we use to evaluate the hypotheses in a test. That is, the test statistic is what we use from the data as evidence against the null hypothesis. The test statistic measures the difference between what we see in the data and what we would expect to see if the null hypothesis, H_0, was true.

Transferring findings   This involves taking the data from one population and transferring the results to another. For example, Londoners' opinions may not be a good indication of the opinions of people from Edinburgh.

Trend   The trend of data on a scatter plot can tell you whether there is a linear (straight-line) relationship, that is, whether you can roughly fit all the dots onto a straight line?

Two-way table of counts   A two-way table of counts, also known as a contingency table, is a tabular representation of the joint distribution of two categorical variables. It displays the frequencies or counts of observations that fall into each combination of categories for the two variables. The rows of the table correspond to one categorical variable, and the columns correspond to another categorical variable. Two-way tables are commonly used in statistical analysis, particularly in the context of chi-squared tests. These tables help researchers and analysts understand the relationship between two categorical variables and assess whether there is a significant association between them.

Validity   This refers to the accuracy of research data. A researcher's data can be said to be valid if the results of the study measurement process are accurate. That is, a measurement instrument is valid to the degree that it measures what it is supposed to measure. There are different types of validity. Internal validity refers to whether there is a causal relationship between the variable being changed (called the independent or explanatory variable) and the variable being measured (called the dependent or response viable). External validity refers to how well one can generalise research results to other settings, programmes, persons, places, etc.

Variable   A variable is a characteristic that can be measured and that can assume different values. Height, age, income, province or country of birth, grades obtained at school and type of housing are all examples of variables. If we were looking at a variable like grades obtained, A and B grade would be examples of different values for that variable. Variables may be classified into two main categories: categorical and numeric.

INDEX

Note: Page numbers followed by "f" represents figures, and "t" represents tables correspondingly.